THE QUANTUM RE-ENCHANTMENT
OF THE REALITY YOU LIVE

Also by Amit Goswami

The Self-Aware Universe

Science and Spirituality

The Visionary Window

Physics of the Soul

The Quantum Doctor

God Is Not Dead

Creative Evolution

How Quantum Activism Can Save Civilization

Quantum Creativity

Quantum Economics

The Everything Answer Book

Quantum Politics

Quantum Activation: Changing Obstacles into Opportunities (with Carl David Blake and Gary Stuart)

See the World as a Five Layered Cake

With Valentina R. Onisor

Quantum Spirituality

The Quantum Brain

The Quantum Re-enchantment of the Reality You Live

HOW QUANTUM SCIENCE IS CHANGING OUR BELIEFS ABOUT MATTER, CONSCIOUSNESS, GOD, AND SOUL

Amit Goswami, PhD
&
Valentina R. Onisor, MD

LUMINARE PRESS

WWW.LUMINAREPRESS.COM

Luminare Press
442 Charnelton St.
Eugene, OR 97401
www.luminarepress.com

LCCN: 2022917502
ISBN: 979-8-88679-088-7

Table of Contents

PROLOGUE

What is enchantment? Children experience it as a special connection to the universe out there. Some verbalize it, some not. As we grow our rational mind, the enchantment gives way to separateness and cultural individuality. For mentally healthy people the enchantment remains in milder form of curiosity about meaning and purpose of life.

The word *spirit* could refer to enchantment; but increasingly it means different things to different people. To some alcohol is the only spirit they recognize; it moves them towards freedom from inhibitions at least.

"How should I live?" This is a question that a fair number of people are still bothered by even in these days of social media. If you are one of them, you will be interested in the title of this book. Obviously, how you should live has much to do with what the nature of reality is. If you want to live what is real, not follow mirages, that is!

So, we start with a formal presentation of what organized thinking tells us about how to live reality these days. Let's take up religions first.

"*Be good. Do good.*" The emissary of religion, a nice-looking guy in ocher robe tells you. He could easily have been a priest, a rabbi, a mullah. You think, Oh, this is the generic religious preacher. But what exactly does he mean? Let me ask.

"Why should I be good, let alone do good? Ok, do good to myself? I try, that makes sense. But to others? Why?"

"That's what the Good Book says! Isn't that enough? Follow."

"Before I follow, I'd like to if you tell me what good it does to me helping another?"

The preacher guy nods. "Good enough. There is a bonus for you too. You go to a nice place called heaven after you die. You know, that is where God lives. And love is everywhere in heaven."

You have heard that before. God's love is everywhere. Not true about earth obviously. Could be true of heaven. Is it worth taking a chance now and change?

You think it over. "Sounds good. But heaven can wait until I am old, I think. Then I will be good to others and go for God's love. But not yet. Now *I* am my only concern."

The preacher fellow is amused by your reaction. "You are in good company. Even the famous St. Augustine felt that way."

But as you turn to leave, the guy calls out. "You know it's not just God's love, although that's the official line."

"There is an unofficial line?" you have never heard that one about the charms of heaven. "This I'd like to know. Now I am curious yellow."

The guy thinks a little. "Well, I am not supposed to gossip about this; it is kind of secret. But…"

"Tell me. It may change my whole life."

"Well, the secret is out any way; at least one Good Book gave the secret away." He still lowers his voice. "Ok, I will tell you. It is really that other kind of love; you know beautiful angelinas are everywhere in heaven and to enjoy it men are given infinite erections."

Well, well. Interesting. Unfortunately, in the age of Viagra it is not that enticing. You remember an older friend telling you how one time he had Viagra hangover and his erection would not go away for four hours. It was actually painful, he said. Your final conclusion, "I will still pass, I think."

You turn around to leave, but the guy interrupts. "Don't be hasty. Ask me what happens if you do harm to others. Because face it! If you are only concentrated on your *numero uno*—you—often you end up hurting another. Right?"

Indeed so, you had to grudgingly admit. You have seen others hurt by your me-centered actions! You are not that callous.

"Ok, tell me why I should *do no harm*?"

"If you do harm, God punishes you. You go a place called hell after you die. Fire and brimstone. Is that where you want to live for eternity? Believe me, you don't want to go there."

"I hear you." You say, not satisfied. "Look, God and heaven make some sense to me. I do believe there is life after death. But all you guys say, God is good. And then in the same breath you say, God punishes you if you do harm. Makes no sense why God should be so inconsistent."

"God punishes you for your own good. So, you learn from your mistakes." The guy says gravely.

"But you said, hell is eternal. If I need to learn from my mistakes, don't I get another chance?"

"Oh yeah, that's our alternative position. It is called reincarnation. We reserve that answer for clever guys like you."

Reincarnation. Intriguing, coming back to life again and again. It may get boring though! The same old nursery rhymes, teen romance, football season, marriage and children. You remember the movie, *Ground hog day*. A news anchorman re-lives a ground hog day over and over until he learns to love. You have to think about that one! You look at your religious guide with some respect but cannot help making a parting jab at him.

"I have to think about that one. by the way, when I was in college, a physics classmate explained to me that according to physics, your God's laws, hell has to be a very cold place, frozen. Not fire and brimstone."

The poor guy looks so puzzled. You are satisfied with yourself. Ready for another view of reality?

For the next presentation, we take you to the classic materialist. There are two guys who represent this view so well that over the years they have become the standard-bearers of this view. One guy is Greek, the other Indian.

The Greek guy first. In answer to your questions, how should I live? What is reality? He says, "What is reality? Why bother? How should you live? Easy, so you can be happy. My recipe is straightforward. Eat, drink, and be merry. Check it out."

You recognize the guy, the philosopher Epicurus. He is prescribing the classic Epicurean delight. Ok, you can relate to that.

But you have always been curious. Why did this guy leave sex out of the equation for happiness?

"You did not include sex. Doesn't sex make us the happiest?"

"It is included; it is part of making merriment. But my friend, you know nothing about happiness, it seems if you think sex is the best. Not after you have discovered philosophy or mathematics, my friend. Sex pales."

You can buy that. You remember the story about the mathematical physicist Richard Feynman whose wife divorced him because "he brings mathematics to bed." Yeah, yeah. This guy Epicurus is right on. You also remember reading about that Playboy cartoon. A guy is reading a book, *The Joy of Sex*, while his naked girlfriend is trying to entice him to sex. The caption says, "Cut it out. Don't you see I am reading?" Indeed, philosophizing about sex can be fun!

Now you are even more curious about what the Indian guy is doing there; after all Epicurus' message is pretty definitive. What can this guy add?

"Yeah, the Greek's message is good. But he misses a very important element of happiness. Actually, a couple. First, when he says 'drink' you think alcohol, right? Don't deny it. But too much alcohol makes you sleepy and druggie, and nonsensical. My advice is ghee."

"What? What is ghee?"

"Clarified unsalted butter."

"That does not sound appetizing."

"Try it with food, like bread, in the beginning." Says Charvaka, introducing himself. "The second thing poor Epicurus misses is money. Where does the money for that happy life style come from? I say, it has to be borrowed. That is the recipe for perpetual happiness. Or else, you go broke."

Wow! This fellow is wise. Also, after a little mulling over, you even begin to see that the advice about ghee may be right on, too. It is like olive oil; you eat salad and bread after soaking with olive oil, as if you are drinking the oil. Ahhh, makes sense. And it is the healthiest diet.

So, have you found your answer to, "How should I live?" We doubt it. The fact is, the materialist recipe loses charm as soon as you add up the amount of time you can spend with food and making merry with sex. What then? Boredom. Philosophy is no challenge to boredom.

So, we take you to the modern preachers of materialism who claim to be scientific.

An academic type was ready with an answer as soon as you finished your question.

"What is reality? An admirable question. But isn't it obvious? Reality is what is a permanent fixture of your life. Like your land, see even our language knows—real estate. Our bodies, the brain, they are real. Mental thoughts, they come and go. Even wise religions like Buddhism admit, thoughts, feelings, their permanence come from your attachment to them. Give up the attachment! And be happy."

Oh, this guy is wise. "How about thoughts like God and heaven? It does sometimes console me thinking that there is a benevolent father figure looking after me!"

"Exactly, God exists only because you give the concept permanence by believing it. Do you watch old episodes of the show *Star Trek*, the original series?" the guy asks.

"Sometimes. Why?"

"We inspired one of the writers to write one episode making this point. Maybe you have seen it. Captain Clark and company meets the Greek God Apollo, whose strength and physical vigor impress the whole crew no end. The God easily gets the captain's girl even; no context until the human guys stop believing in the God. The God disappears in a mist as soon as people stop believing. See, this is what we are doing now."

"You are debunking God. I have heard about that. But there are still so many believers!"

"You have to remember. The God-belief is a very old one. It will take some time. We have a plan."

It is obvious that you are all ears, ready to listen. The guy continues:

"See, we first started with Nietzsche. With a catchy slogan. God is dead. All power is in the physical laws. God as a father figure comforting us, helping us in need is useless when it is clear that at best God makes the laws and then retires.

At the next stage, we attacked the idea that God creates life. The time was ripe when we discovered the double helix structure of the DNA. Now that is a powerful image to convince people that life is simply the play of complex molecules like DNA with double helix and the proteins that can fold in a zillion mysterious ways.

At the third step, we threw in the idea of artificial intelligence. Our intelligence did not come from God; it came from evolution out of survival motif; we are simply machines and our brain is a cognizing computer."

Oh, this guy can speak powerfully, he has a good brain, a computer that knows.

The guy continues; he knows he is getting through to you. "See, how it works. The brain makes a lot of noise, electrical noise; it's an electromagnetic machine you know. Then it makes sense out of its own noise hallucinations and fools itself with concepts such as God and love. Those concepts are unreal, illusory ideas, not to be taken seriously."

Being a guy, you can't help your concern about women and their claim that only love can make you happy and love is the way to live. To you, love is sex, and sex is good. But it is only seven minutes. Even with Viagra, it is no more than half hour. And when you do it that long it gets boring anyway. Not particularly happy. The end, orgasm is great; but it is just that few seconds of ecstasy. Then wait for the next bout. And women constantly claiming to be sore…

You did not even realize you were voicing your complaints. The scientific materialist was sympathetic with you plight with women.

"Yeah, we invented birth control pills so that women can join us in the enthusiasm for the quest for unlimited sex, and see what happens? They found the women's lib movement! They started talking women's values. Women's values, my foot! What ingrates! Anyway, we think we have vanquished the women's libbers; we have coopted them."

"You were explaining the steps for loosening people's belief in God. What next?"

"We gave you high tech and the concept that information is everything; information is the way the world ticks. Also, it gives your brain a rest from all that cognizing—meaning processing. Information processing is more mechanical.

"Look! How happy people are with trivial pursuits and social media. No more boredom. No boredom is no-more-dom. You don't need anything else to be happy."

You object. "You are simplifying. I have been there, done that. It does not satisfy for long. It itself gets boring. Rest or not, I need my brain to cognize. You comprehend?"

"We have an option for people like you.

"We have created Donald Trump."

You don't get that one. You thought it is Democrats who believe in science, the materialist science that this guy preaches; and Trump is anti-science. Why would scientists create him?

"You what?"

"You heard me. The problem with Democrats has been that they are materialists but they feel compelled to support moral values, even though they know that the values have to be all a pretend. Trump showed that people hate pretend values; now people can be authentic in their hate. Trump is the ideal perfect materialist; me-centered to the utmost, uses negative emotions to manipulate people and gets his way; sings the virtues of using people for pleasure and does it too. And guess what? Women love him anyway! Democrats will see the light and come around. There is no shame being authentic to what we really are: machines with cognition that can recognize what it needs to survive the best with built-in negativity to achieve what it wants including its pleasures."

"So, you are saying, morality—be good, do good—is humbug?"

"Exactly. Matter is the only reality. How should you live? You live to maximize your survival and that means pleasure, doing whatever it takes."

As you start to leave, the guy speaks to your back. "Hey that didn't sound so good, did it? Talk to that guy over there. He will give a positive spin."

So, we take you to another fellow who seems to be computer geek. He hears you with sympathy. "Yeah, what if you are not a leader, but are on the other side, a follower. While guys like Trump exists, the followers do okay; they bask in their leader's glory; they live the leader's life vicariously. Our brains are wonderful weaving illusions from its noise. But what if there is no Trump in the horizon?"

"Yeah. What then?"

"The next stage. Robotics. With robots, everyone can build a harem of followers and simulate Trump within himself, even herself."

You are not totally sure how this will be economically feasible. However, economics is your weak point.

Your guide sees your hesitation to approve. "You don't get it. This is in the future any way. In the meantime, that guy in the ponytail has another way for the ordinary guy to live optimally with reality.

So, we take you to a pony-tailed guy with a lapel sign that says Positive psychology. You know about them; a guy from Harvard initiated it and now has quite a large following.

You ask him point blank, no pussy-footing, "How should I live? What's your point?"

The guy gives you an understanding smile. "I know. Facing reality is so hard for the brain. Sometimes it probably wishes it were a dumb machine and couldn't cognize. See the problem! Animals live for survival; the selfish genes drive their lives and it is simple. They eat and have sex to live which is survival. Then natural selection created the brain, created its pleasure circuits, created molecules of pleasure. For humans, the motto of living is pleasure: humans live to eat and have sex.

But the negative emotions that were good for survival are of little use for pleasure. Ok. Some folks' special brains use the negative to dominate and others to serve them for pleasure. But only few brains can do that. That is why your disappointment! What do you do if you are not a hunk or a jock but a nerd?

You nod thoughtfully. Good point. How do you get a girl to date you when every girl wants a jock? And if you are a girl, how do you get a jock to date you if every girl competes with you for it? These were the perpetual problems at high school.

The ponytail continues. "Nature again. This time it is epigenetics. The brain cells epigenetically reprogram the brain's pleasure circuits for territoriality not only to defend its territory but to expand it.

It is a fact that if you don't use your eyes, like blind people, the optical sensory part of the brain starts losing its territory to the hearing sensors of the brain. So, the brain's strategy (which you experience as yours) is to keep the negative circuits as unoccupied as possible. How? When a negative stimulus comes in, give it a positive spin. Act positively."

"Isn't that pretension?" you are skeptical.

"True, not all brains can do it. Fortunately, the brain has all that noise to make belief."

You are still sceptic. "Aren't you putting a lot of human characteristic into the machine, like intention?"

"Not any old machine, pleasure-seeking survival machines that can cognize. It's a lot of power packed in this three-pound universe." He proudly taps his brain.

Can the brain be all that without any help? This is the question that can free you from the materialist delusion. That and the question, there are so many transformed people everywhere at all ages. How do they get there?

Well, they practice and explore the archetypal values; that is an open book to read. Religions, as archaic as their beliefs may sound—hell, heaven, even God—are not to be rejected out of hand. Perhaps there is more meaning in religion that meets the eye at first?

You live in the age of worldview polarization, no denying that. Both worldviews have truth, both have dogmas. Both are elitists; they want to keep you in the same place so the leaders can dominate you.

The polarization has given us the current crisis, the most recent ones are breakdown of capitalism and democracy. Crisis also means opportunity. Opportunity to examine the case of reality, WHAT IS REALLY REAL?

Religions did not come from vacuum; they came from rationalization of the teachings of great transformed people, whom humanity has cared to remember to this day, in some cases, for millennia. These teachings—the basic theme is enchantment of matter via the infusion of a Spiritual Oneness—are surprisingly still very alive among smaller groups of people. They are said to belong to the esoteric branch of religions. Together let's call them people of the spiritual wisdom tradition; scientists use the slightly pejorative term mysticism, but never mind. We will use that popular term too.

These people taught spiritual wisdom of enchantment on the basis of their theories, no doubt; however, their theories were always based on direct experiences. And they themselves followed through with transformation that others could and did witness. Their modus operandi was to use the two prongs—theory and experience—in tandem to explore truth. Much like science until science became dogmatic, do you see?

Synchronistically, there is an aborning new worldview unfolding via our exploration of the newest physics paradigm on the block—quantum physics. What is truly amazing is that the emerging ideas of quantum physics are converging to the same metaphysical ideas of the spiritual wisdom traditions. Some people, maybe as large as 15% of humanity, are excited; they are talking about the integration of science and spirituality. Does this idea appeal to you?

The point to see here for you to continue reading this book is this: spirit and matter are not exclusive. *Spirit cannot express without matter; matter cannot be enchanted without the spirit.*

With the acceptance of the materialist dogma, science has rejected the spirit. Popular religions have compromised the spirit in their vain search to dominate ordinary people. In this way have you, a majority of us, become this dispirited, unhappy, confused people.

Can the new quantum worldview lead to an integration of spirit into science leading to a re-enchantment of reality we embody? In the following pages, let's explore the question together, shall we?

CHAPTER 1

Introduction: Philosophy, Darshan, Science, and Dogmas

Philosophy

If you ask the proverbial "man on the street," "what is reality?" he will be a bit puzzled. "That is a question for the philosopher. Why ask me?"

In philosophy, BEING is where the question of reality begins. Philosophers call the subject ontology or metaphysics. Becoming is about changes of reality over time, such as evolution and development. Evolution and developmental history from being to becoming is called cosmology.

Then there is what philosophers call epistemology—how to know Being. Ways of knowing reality. Most importantly, *philosophy is thinking about reality rationally.* Ontology, cosmology, epistemology, everything.

One of the first things about reality that we all notice is that we have external experiences of matter *as well as internal experiences of mind, soul, and spirit.* The two experiences are qualitatively different. So, philosophers everywhere developed the philosophy of dualism—internal and external belong to two separate and independent domains of reality. Rationality dictates that. How do the two worlds—domains—interact though? This question was never adequately answered. Two separate domains of reality have nothing in common; they need a mediator to communicate. The question of mediator is a difficult one.

Darshan is a Sanskrit word meaning philosophy based on trans-rational experiences that reveals new insight. Creative theorizing—

theory making based on discontinuous creative experiences rather than step-by-step continuous rational thinking—is the modern name for darshan. Darshan—creative experiences are beyond the rational mind, also beyond our ordinary experiences.

Darshan—the elevating philosophy of the wisdom traditions—has given us ideas about consciousness—archetypes or epitomes of thinking that we value, that have served as the basis of human civilization world over. It has given us the concept of oneness of consciousness that immediately leads to the fundamental ethic: do no harm.

How far does the oneness go? As far as possible. Lest there is any confusion, darshan theorists define their ontology as nondualism—Oneness is One without a second. The ontology has also been called monistic idealism.

Oneness sends the archetypal values to us in material manifestation via the subtle vital and mental bodies. The soul is made of the representation of the archetypes in the vital and mental bodies—positive emotions. We experience the movements of the vital as feeling; feeling plus thought make emotion.

What is matter though? Put crudely, can matter be made of consciousness? Naïve thinkers really say (the paucity of good philosophy!) this: may be thought condenses into matter just like water vapor condenses into liquid water. But they really are missing a point, a very important point of distinction between consciousness and its experiences so far in our discussion. We will discuss the point later.

In any case, ordinary people do not experience reality that way as One without a second, do they? They see themselves as separate from each other. Explanation: Oneness is beyond rational thinking, beyond ordinary sensory way of looking.

Darshan laid out an epistemology as well. Meditate. Over millennia, many forms of meditation were suggested by wisdom traditions. Today, they are indeed found helpful for the creative process. So, there is renewed interest in meditation.

However, people of ordinary human condition cannot meditate, they get bored. Besides, the idea of employing the nonrational for philosophizing or theorizing about reality did not sit well with the pervading rationalism of human culture of ordinary people in all cultures. In this way developed rational apologies of Darshan called religions. The idea of Oneness degenerated into the idea of God, our Lord and Master, King of kings. God lives in Oneness—heaven, we live in separateness, earth. Dualism again; it made a comeback by popular demand and because the followers of the great seers of darshan in leadership position did not like to meditate to find out if it is possible to go beyond rational thought.

In trying to rationalize away aspects of reality, religions created dogmas—ideas that were declared infallible. They have to be taken on faith. How does God interact with the world? Don't ask; God does, take it on faith.

How is faith different from a belief? Why can't faith be challenged? Silence. Soon other dogmas were added, different from one group to another. Instead of one wisdom tradition, we now had a smorgasbord of religions; each excluded the others.

Subsequently, people found that the application of the ideas of esoteric darshan or popular religions to worldly affairs of humans do not always work, causing confusion. How do we live if neither the wisdom traditions nor the rational version—religions—are giving us adequate guidance?

In response, there came traditional science of matter, knowing material reality with the combination of creative theory plus experiment—collecting objective data—data about which we can get general consensus.

Isaac Newton developed the science of macroscopic objects of bulk summarized in a mathematical equation. This equation gives us a deterministic predictable world that seems to be mechanical and controllable.

What works for inanimate matter, can it work for us living creatures? A scientist named Charles Darwin gave us a theory of life's evolution based on an idea from human behavior applied to rudimentary life. The idea is survival. Except for the idea of survival, the theory did not invoke any other nonmaterial idea. Rationalists jumped into the conclusion, philosophy again, not science: Darwinism is a triumph of materialism. All is matter. Do molecules of matter try to survive? Nobody asked at the time. No one answered.

Subsequently, researchers discovered programmed macro molecules—DNA, RNA, and protein—that are exclusive to life.

At about the same time, computer scientists claimed that human intelligence can be explained as derivable from machine intelligence when sufficiently powerful computing machines were developed.

In a few decades, scientific rationalists declared: all is rational, all is matter and mechanical. Science acquired a dogma. The same infallibility and exclusion as religions befell it. Science became materialist science.

The result was worldview polarization that we see today between materialist science and religion. Being exclusive of one another, the followers fight creating chaos. The situation is this: materialist science has done good for humanity in the material arena; religions have done good in affairs where consciousness is needed. Science excludes consciousness and archetypal values. Religions want to take us back to old-fashioned feudal societies that cannot handle technological progress that has undoubtedly made human life better.

The newest science on the block is quantum science—a mathematical science developed for explaining non-deterministic behavior of elementary particles of matter and energy (elementary particles of energy are called quantum, hence the name). This mathematics is based on creative theory and experimental data, but to understand it, we have to include something new and unexpected—consciousness and experience. This has given us the hope of integrating the two most successful methods of knowing reality—darshan and science without dogma.

Why do things have to be so complicated, so controversial to grasp? The limitation is entirely ours. Reality comes to us through our experience. Among our experiences, rational thinking dominates. We make models to sort out our experiences mostly with rationalism, few with nonrational experiences. At any given time, our models depend on our ability to experience, experiment, and rationalize. Models change because our abilities and how we use them change over time.

This is why getting a historical perspective paying special attention to the prevalent society and culture of a particular model is important. This is the objective of this book.

Human Experience

The first undisputable striking thing about our experience is that some of it seems external to us, some of it internal.

The external part is public, and can be subjected to building a consensus about it. You see a chair, I can see it too, and pretty much the same way so that we can reach a consensus about it: there is a chair over there.

However, some of our experiences seem internal and private. For example, thoughts and emotions. Good that it is this way, no? We think such weird things, some time; it would be embarrassing to think that somebody is reading my mind at those moments.

So historically, the very first models of reality reflected this inner-outer dichotomy of our experiences. The model is the above-mentioned dualism: dual domains of reality. In current language, we call the objects of external reality matter; the material reality—matter moving in space and time—define the physical reality. In contrast, dualists assume that the objects of internal reality are nonphysical.

As mentioned before, the obvious problem for this model is the question of interaction or communication between the two domains of reality. Physical and nonphysical cannot interact without a mediator. Dualists never could answer the question of the mediator.

Reality has to be monistic—made of one thing. One such model is called material monism. There is just one reality; obviously, the one that nobody can deny, the one external, consensus reality of matter moving in space and time. We have different view about things of our experience—subjectivity is the problem. This model declares: reality should be objective, why fight about it? In this way, another popular name for material monism is physical realism. Everything real is physical.

How do we explain the internal experiences then? That is easy. The internal is epiphenomena—a somewhat derogatory word meaning secondary phenomena—of matter. The internal is ornamental detail.

A neuroscientist famously said when asked, "What is mind?" He said, "Doesn't matter." He had a certain wit though. When the questioner asked, "What's matter?" The neuroscientist replied, "Never mind."

What gets lost in this very public polarized debate is the other monistic candidate for reality—Darshan or spiritual wisdom tradition propounds monistic idealism. Everything is consciousness, even matter.

The newest physics on the block—quantum physics—discovered when the behavior of submicroscopic elementary particles of energy and matter was explored, shows that matter may not be the way we experience bulk matter—solid, rigid, etc. The elementary objects of energy are called quanta (plural of quantum), hence the name quantum physics of the dynamics.

Quantum equation for predicting the movement of the elementary objects of matter shows unequivocally that these objects are waves, they spread out and not travel in determined trajectories (fig. 1).

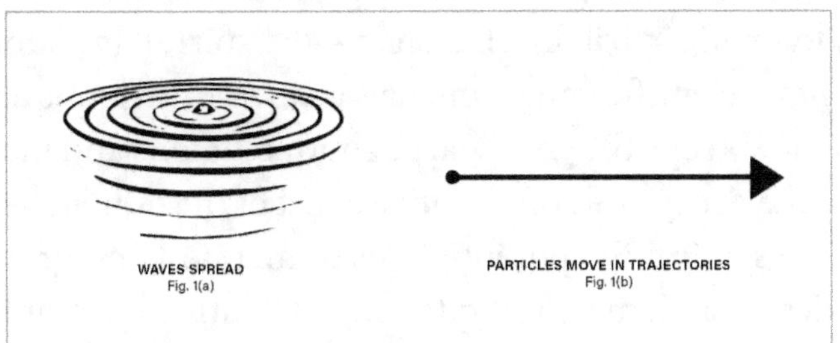

Figure 1 a) Waves spread out.
Figure 1 b) A particle can be only at one place at one time.
It travels in a trajectory.

This would be a paradox (the same object cannot be a wave and a particle at the same time) except that they belong to a different domain of reality outside space and time; they are waves of possibility. Only when we measure, they become particle.

Get this. An electron as a wave of possibility moves as expanding crests that embrace many possible positions at the same time; nothing like the rigid solid material object that you see and more like your internal thoughts that change inexplicably to many different thoughts whenever you lose focus. In this way, could matter be ideas of consciousness; could it be that consciousness chooses the actuality from the many possibilities of the electron's position? Yes, said many quantum physicists beginning with John von Neuman. Eventually, Amit found the paradox-free way of resolving the whole situation and declared the final unambiguous meaning of quantum physics: Oneness consciousness is the ground of being and matter consists of possibilities of consciousness to choose from. When consciousness chooses, the wave (many possible facets) collapse to a particle (one actual facet).

We have to reckon with the new reality: physics, physical reality includes both material and the nonmaterial with nonmaterial consciousness being the basis. The model of reality that postulates everything is matter must be called *material* realism, not physical realism. Nonmaterial mind and soul are as real as matter; all are possibilities of consciousness to choose from.

Literally, quantum physics is integrating science as per its original nondogmatic spirit and Darshan—the spiritual wisdom traditions. A new scientific paradigm, nay, a new worldview is aborning. The new view keeps the good part of both religions and materialist science and yet captures the creative spirit of the human being. We are creatures of infinite potentialities to manifest and grow.

The details are amazing. In the new scientific view, at the base level, a human being is dominated by rational and me-centered thoughts, negative emotions, and pleasure seeking; as we explore the potentialities of values—the archetypes—we develop the capacity of positive emotions in our repertoire of regular experiences; this latter is the beginning of soul-making. The end of the journey—the destination—is defined by what we call spirit. *Want to join the adventure in soul-making?*

Before you begin the journey, let us caution you: we live in a time of intellectual anarchy. Materialism in the original form cannot explain quantum physics; but so-called post-materialists continue to propose seemingly avant-garde integrative ideas to co-opt spirituality without really breaking free from the matter-is-everything belief. These people have mastered the art of sophistry creating a very confusing situation that materialists can simply ignore.

How can tell these intellectual masturbations from the real thing? Only by grasping the hard question, what is consciousness?

What is Consciousness?

What is consciousness? Don't be surprised if you cannot explain the concept right away. Why? You cannot experience consciousness as separate from yourself like you can a material object or even a mental object.

Hidden in the label of internal, there is this one experience which is qualitatively different from all the other experiences: material objects, thoughts, emotions. There is something common about material objects, thoughts and emotions, even archetypes, they all are objects.

Behold! There is also a seemingly internal experience larking here, kind of implicit, hard to notice. For a long time, people did not much notice it, because they experienced it differently than today. Today, and this goes back millennia, language gives it away. When we describe an experience, we use first person I, "I am singing a song." There are two poles of the experience: I is the subject and the song is the object. Experience, or the awareness—the internal space that embodies the experience—has two poles: subject and object.

Now that we are putting it this way, it should be quite clear. An experience must have an experiencer—the subject.

In Sanskrit, I is *aham*. Some seven thousand years ago, some researchers in India was asking the question, "What is I?" Or linguistically more appropriately, "Who am I?" Because, even a little research, they called it meditation, showed them, that this experiencing I before meditation, is quite constricted, centered on me, myself, the individual that I can also experience as an object. In Sanskrit, object—it—is translated as *idam*; the constricted consciousness they called *Idamasmita*, I-am-this-ness. But then meditating regularly results in a different experience of the "I." After a meditation, we experience a relaxed, expanded consciousness, that is generous, that can include another in its care. Consciousness of this I is no longer an object or me; it is distinct, its distinction is our I-ness, *asmita* in Sanskrit. So, the question, "Who am I?" The constricted me-centered I that in modern times we call ego, or the expanded consciousness asmita?

Then, something unforeseen took place in the experience of some of these researchers. They fell into a state in which their "I" seemed to engulf all other people, creatures, and things. They called this Oneness *Brahman* in Sanskrit, which today translates as consciousness.

These seers founded Darshan, philosophy of experience. Ultimate reality is Oneness consciousness. The Oneness splits into an experiencing self or subject and the object of an experience by the action of a "force" that produces the illusion of separation. The Sanskrit word for illusion is *maya*; accordingly, the force was called maya as well.

In one interpretation of this philosophy of Darshan, Oneness alone is reality, all duality is illusory. In this way, this philosophy is called non-dualism.

However, in another interpretation, the illusion is a necessary ingredient to produce experience. Experience is for consciousness to explore itself, its ideas. It is a necessary part of reality. Since consciousness and its ideas are the basis of reality, and in the ultimate reckoning, they are one, this philosophy is called *qualified nondualism*. Both fall into the category of monistic idealism as opposed to material realism.

But of course, this idealist alternative to dualism that a part of the internal experience—the self—is the representation of the real reality, does not sit well with rationalists, especially for people in the West. The philosopher René Descartes set the trend: *I think, therefore I am (me)*. In contrast, the Easterners seem to be saying, *I don't think (of me), therefore I am*. This became the East-West divide that still influences a lot of people's thinking.

God

And then there is this overarching concept of God that still dominates a lot of discussion about reality. What is God? Does God exist?

The concept of God with a capital G evolved in steps. First, erudite people among our ancestors who were hunters and gatherers noticed that there are a lot of natural phenomena for which they could not find any material explanation, for example, thunder and lightning, even rain. So, they invented the concept of nature god with small g to explain such natural phenomena. What is the explanation of rain? Rain god makes rain. Like that.

The hunters and gathers were survival oriented, mainly concerned about the material reality around them—food to hunt and food to gather. Thar archetypes of importance were power; hunting needed that; and abundance, both hunting and gathering were focused on that. Both hunters and gatherers needed relaxation; the explorers of the archetype of beauty provided it—the musicians and the artists.

Later, the advent of civilization produced the capacity of garden agriculture and families grew around that. Men and women working together discovered the archetype of love. How the leader of a community dealt with the members led to the discovery of another archetype—justice. Soon, the concept of nature gods was replaced by archetypal gods.

Unfortunately, different communities chose different gods to explore or worship causing division. So, the erudites combined all the little gods as subordinates to one almighty God of capital G, the ruler of the universe, all reality. In this way, monotheistic religions cropped up everywhere.

Note that this is also the philosophy of dualism in another garb. Instead of mind-matter dualism, this is God-world dualism. The world mostly runs with material power, no doubt; but God represents a higher power. That miracles—phenomena without explanation via material cause—happen is a proof of this higher power, according to this view of reality.

But of course, critiques can raise the same question of interaction and mediation against God-world dualism as well. Obviously, God has to be nonmaterial being the source of nonmaterial cause. How does God interact with the world, with humans then? Where is the mediator?

But this time, the proponents of dualism, the priests of religions, were better equipped conceptually. God is almighty, a higher power, can do anything. Call it miracle when God intervenes with the world. You have to take it on faith.

How Spiritual Wisdom Traditions Adapted Religion

In the olden times, there were not much communication between different communities of the world. The transition from the concept of many small g gods to one God of monotheism was not universal.

In India, we find the idea of nature gods giving way to the idea of archetypal gods. And then, about seven thousand years ago, well after the discovery of large-scale agriculture with ploughs, monistic idealism was discovered by people who experienced what they propounded and their moral authority was quick to establish itself.

The savants did not exactly throw away the previous knowledge and useful rituals to explore the archetypes. They found room for the archetypes in the somewhat rarefied experience of intuition. In this way, Hinduism never gave up the idea of small g gods. Hinduism additionally found room for big G God as well—God (Sanskrit *Ishwara)* is primal principle; all little g gods are attributes of it,; it is also called the male-principle purusha as opposed to the female-principle prakriti (Sanskrit for nature). Manifestation is the product of the joining of purusha and prakriti. So, the wisdom went. This is called *Samkhya* version of God-world dualism.

The great masters—the seers of darshan—brought back the older dualistic philosophy of Samkhya of basically God-world dualism and recast it in the Vedantic picture. Consciousness is the ground of all being. In order to know itself, it has to create limitations, so its involutes (via the action of maya of course), introduces qualifications as a precursor to the eventual subject-object duality. The oneness thus splits into a universal subject (*purusha* or *Ishvara*, akin to God), and a universal object (prakriti). Only then, maya acts for one more time to produce the subject-object split of experience. This philosophy is what we called qualified nondualism before in a slightly different garb. Maya's action continues and produces further forgetfulness to the state of everyday experience—the ego.

In the Middle East things happened differently. They're developed Judaism, which eventually produced two offspring: Christianity and Islam. All three are sometimes collectively called Judeo-Christian religions. In the original Judaism, monotheism was regarded as a high accomplishment, the role of the little gods is somewhat deemphasized, they are called angels—messengers of God. God in these religions is a strict disciplinarian, and you are punished if you violet God's commandments.

But of course, Christianity was famously founded by Jesus the Christ touted later as "only begotten son of God", and Islam by "God's messenger" Mohammad. These two people were genuine seers in the same wisdom tradition as in Hinduism. They propounded their wisdom in the same primacy of Oneness way, God is Oneness. They emphasized the archetypes, especially love.

Later followers were not all seers; in fact, non-seers—rationalizers—were more influential than seers. In their non-seers' interpretation, both of these religions retained the punishment aspect of the older religion. Punishment and sin are easier way to manage the parishioners than the positive message of love.

In this way, in stead of Oneness and consciousness, religions world over became identified with God-world dualism and became the subject of serious controversy.

There is something here to be noted though. Most, if not all religions, do attract a few people who are interested in the nature of reality and what that implies for human life. These people very quickly discover a whole bunch of previous explorers who were similarly interested, and they continue the wisdom tradition. In India, this continuity of the wisdom tradition is called *guru parampara*—Sanskrit that translates as "there are always seers." But this is true of all religious traditions. In this way, religions generally have an esoteric tradition apart from the popular and very public exoteric tradition.

Back to Objects of Experience: Gross and Subtle Bodies

Let's get back to the discussion of the object-pole of experiences. The material sensing experience dominates us no doubt along with thinking. But of course, who can deny emotions? Under their throw, we behave differently. There is passion, there is energy in emotions.

In earlier times, when people looked at their experiences without preconceptions, they could see the energy aspect of emotions. In this way, first India and then China developed the concept of vital energy—energy connected with living. Indians called it by the Sanskrit word *prana* which also means life. Chinese by the Chinese word chi. Just as we have the physical body with which we sense, they conceptualized a vital body with which we feel the vital energy.

Similarly, the Hindus conceptualized a mental body with which we think.

How about the archetypes? Indeed, in the Vedanta literature itself, we see reference to a body from which *vi-jnana*—context of knowledge—comes to us. Obviously, they were conceptualizing the experience that today we call intuition, the objects of which are the archetypes.

Amazingly, the Jewish esoteric tradition has their own Vedanta which is called the Kabbala and it, too, has a reference to four kinds of bodies. These bodies are embedded within the spirit—consciousness—of course, the fifth body.

In this way, we have consciousness which is called the *causal body*; the nonmaterial cause—maya—comes from here. Then there is the material body that all can sense—the *gross body*. The rest—the vital body, the mental body, and the archetypal body—they constitute the *subtle body*.

Note: Beware of a *misconception* that propagates so you can eradicate it from your belief system: the model that we are made of five elements—earth, water, air, fire, and ether. Replace that unscientific idea with the idea of the five bodies above. Now see that the earlier model was metaphorically quite sensible: earth—solid—represents the gross material body. Water—a liquid—represent the subtle vital; and air—gaseous and subtler—represents the mental. Fire—transformational—represents the archetypal. Finally, ether—all pervading—represents consciousness.

There was also the idea that when the gross body dies, the subtle and the spirit bodies survive. The spirit is the same for all of us; but the subtle carries our individuality. This surviving subtle body reincarnates.

Unfortunately, politics entered the Christian movement and in 5th century AD, the idea of reincarnation was banished from Christianity, although survival of the subtle body after death was still accepted plus the idea of a long wait in purgatory, if needed. The implicit understanding may be that the stay in purgatory is the collective name for all the reincarnations a surviving subtle body takes on before it attains liberation to heaven.

Religions and Archetypes: Spiritual Values

All religions agree about one other thing—moral and spiritual values. What is archetype in esoteric wisdom tradition became religious value, call it virtue, the pursuit of which earned you merits for heaven after death. Violating the values is sin—demerits that sends you to hell.

However, connecting moral or ethical values to a reward-punishment behavioral schema produced a morally rigid society everywhere. Ethics or morality is primarily about being good to others. Of course, humanity has always known about our evil tendencies. The original spiritual idea was to explore goodness to balance and eventually transcend these evil tendencies in the human condition. Unfortunately, religions did a number on this.

Dualism hit again. Religions split the archetype of goodness into a dichotomy of good and evil. And more. Goodness is a primary attribute of Oneness. This is obvious. But when religions personify oneness as God, what then? God has to be only good, right? So, another persona is created for evil; call it devil, *satan*, *danava* (in Sanskrit), whatever. But of course, most of evil is built into *us* as negative emotional brain circuits as modern science has discovered.

Let's get back to being good to combat our evil tendencies. Indeed, being good to women does preclude sexual harassment and abuse of women by men. However, between consensual partners, why should sex for pleasure not be allowed? Yet, the idea that sex for pleasure is morally unacceptable was pushed by the Christian church.

This led to Victorian morality of such a sexually depraved society that eventually when with the invention of the birth-control pill and women's lib movement, sex-for-pleasure found acceptance in society, it went wildly swinging the other way producing a permissive society that borders on hedonism. This certainly facilitated the advent of materialist dogma in science. Materialist science makes pleasure virtually the only positive experience we have.

Even with such adverse reaction, and even after widespread sexual scandals among religious leaders, Catholic priest are not even allowed to marry to this day. This all has hurt the image of religion (and by inference, spirituality) in the West.

The truth is this: sex is potentially a path to the exploration of love between intimate people. Religions have supported marriage and committed relationship between two people; and yet missing the love angle of sexuality, religions have missed out on a major path for the exploration of love.

On the good side, up until the rise of material monism in science, religions, especially Christianity in the West have greatly helped maintain the moral and ethical element in society. This is a key reason for capitalism's success. Please notice that after ethics and morality became ambiguous with the advent of materialist science and science/religion worldview polarization, capitalism no longer works.

Having acknowledging that, we must also note that the religions overall was a degradation of the spirit of spirituality. This is not only the compromise of the metaphysical oneness replacing it by the concept of God but also the compromise on the meaning and purpose of life aspect of the archetypes. The reward and punishment method may have succeeded in putting more people in the fold and may have prevented major mischief that the negativity of our brain might have led to, but overall it was a rather pessimistic way of life that religions prescribed. No wonder, this led to the dark ages in the West. In India, the docile nature of the religious society made India week against barbarian invasions.

The role of the archetypes is in part ethics and morality, no doubt but this covers just two of the archetypes—love and goodness. There are seven others: abundance, power, truth, beauty, justice, wholeness, and self.

Religions have neglected them all. Of these dalliances, the negligence of truth has been the most disastrous. Truth has to be discovered and then lived. Ok, religions assume that their founder had already discovered the ultimate truth; all they need to do is to follow. Unfortunately, another person's truth, even from a great master does not do it for the disciples. "Truth is a pathless land" in the final reckoning. When we try to understand the truth of a good book or a Master's teaching via rational thinking, we interpret, and miss it.

Moreover, one has to discover truth not only to proclaim it but also to live it. Truth not lived does not inspire anyone, yourself included. Religious leaders have been remarkably short of walking their talk all through history, in all cultures.

Religions have also been remarkably short on the archetype of justice. Witness the caste system in Hinduism, the homophobia of Christianity, the male domination of virtually all religions.

The divide between spirituality and religion later became the divide between religion and science with disastrous consequence. Another divide that contributed to the overall confusion of the exploration of reality is the East-West divide.

The East-West Divide

In the sixteenth century West, there was the very influential philosopher René Descartes. Among other things, he said, I think therefore I am. He identified the self with the mind. He also created mind-matter dualism and for him our internal experience consisted of nothing but thought. In other words, psyche = mind and that includes the self.

Descartes' book is called *Meditations* but here again the implied meaning of the word is contemplation in the form of rational thinking.

Ever since, Western thinking has centered on rationality. Only in the nineteen nineties did Western researchers of the brain acknowledged Descartes' error and allowed emotions to be a legitimate experience of the psyche. But the damage was already done.

One of these damages is the East-West divide. When Europeans had the technology and weapons supremacy over the rest of the world, they ventured out and colonized India, the crown jewel of the East. And there they found the very intriguing discourses of Vedanta and Hinduism. They could not understand darshan. What else could it be but philosophy? Wrong philosophy at that. Who can understand the East, they think differently. The poet Rudyard Kipling captured this sentiment perfectly:

East is East, and West is West
The twain shall never meet.

Add to this the fact that the West was the conqueror. The dominator always thinks of the dominated as inferior people. The philosophy of inferior people must be inferior, not worth engaging.

In this way, Western researchers of consciousness got bogged down into rational philosophy. This continues even today. In fact, it is worse. With science came the most powerful vehicle of rational thinking—mathematics. Physics of matter is mathematical, this is the law.

Even today, people ask, what is the mathematical equation for consciousness? Without mathematics how can we trust a science of consciousness? Obviously, they are thinking of consciousness as an object a la Descartes.

Philosophers of science theorized about how science is done, how people find the right equation. They formulated the scientific methods, the step-by-step algorithm for doing science:

1. Think and try an equation that seems appropriate;

2. Deduce the measurable consequence of the equation; in other word make predictions;

3. Verify, you or somebody else can do it too. If the data does not agree with your prediction, go back to the drawing board.

4. Start again. Sooner or later you will succeed.

Fortunately, in the past hundred years, field researchers have really done exhaustive study how scientists, both mathematical scientists (physicists) and nonmathematical scientists—biologists for example, and empirically discovered that scientific models are discovered by a process that goes like this:

1. Preparation. Yes, the starting point is the same; you try various ideas. If mathematics is appropriate, try mathematical ideas as well.

2. Relax, do nothing. Just be.

3. Suddenly, quite discontinuously, not by rational step-by-step, an idea will pop up. Call it sudden discontinuous insight.

4. Now develop the theory manifesting this insightful idea. That is give the idea form. If the form is amenable to mathematics, fine; if not, so be it.

These researchers are giving us a nonrational process they call the creative process. They are claiming that science is not done by rational thinking alone.

First, empirical data of the brain led to the acceptance of emotions. Second, discovery of creativity showed the limits of rational thinking. Finally, there is some chink in the armor of rationalists who still talk proudly about Western civilization. Pride is okay. But the pride does not belong just to the West. Easterners have been using creativity for consciousness research for seven thousand years!

In truth, creativity has been used for consciousness research also in the West by esoteric Christians such as St. Francis of Assisi and St. Theresa of Avila. They just have not been acknowledged. It is time to acknowledge them. It is also time to acknowledge the creativity of Newton, Darwin, Freud, Einstein, Heisenberg, and Schrödinger.

And as we do that, ask, why hold on to a useless divide that limits human experience in a major way?

Science-Religion Divide

The most serious divide today, the scourge of current times is the science-religion divide that have so influenced people that people in different camps have begun to see the world differently. In other words, people are divided into two worldviews. One based on science, the other based on religion.

We have already pointed out that religions brought dogma in spiritual inquiries and limited their ability to inquire freely. The dogmatism of religion kept religious leaders from accepting the free truth-seeking spirit of scientists when modern science began its journey four hundred years ago.

Descartes did something good here in suggesting a truce: let science handle inanimate matter; let religions handle God's territory—the human mind. This truth in some reckoning defines the philosophy of modernism.

Biology is kind of borderline. Actuality, Christianity has traditionally held that animals are machines; so, the initial forays of science in trying to study living creatures was deemed ok by the Church. But all hell broke loose when Darwin theorized that humans evolve from monkeys and in fact, backed up his theory with fossil data.

Unfortunately, this conflicted with the Book of Genesis where it is part of Gospel truth that God created the world and all living creatures in six days, only 6000 years ago. Well, the dates are not as important. But to give up the idea that God created life! To accept that human evolved from monkeys! Impossible. The Christian Church has not changed its position, and this continues to be an ongoing rift.

Scientists on the other hand were encouraged to pursue the idea that the materialist science that works for physics with very little extension can explain life's evolution, if not life itself. The little idea is survival; maybe there are molecules that are driven by survivability. These assemblies of surviving molecular configurations are what defines life. Scientists theorized initially only tentatively. This is ok.

This hypothesis is called molecular biology: Biology = chemistry = physics. Unfortunately, around the nineteen sixties and seventies this hypothesis became dogma. Has the hypothesis been verified? No, but it will be in the future, asserts the biological establishment.

And guess what? Another success came about at about the same time—the nineteen fifties—when molecular biology was taking shape. Computers were invented and immediately people started applying it to produce artificial intelligence. Maybe human beings can be demonstrated to be no more than a machine by building computers that can think like humans! Promissory, but that's ok by the spirit of science.

Meanwhile high energy physicists promised that they are very close to discover the ultimate theory of everything.

What broke the proverbial camel's back is the work of rational philosophers again—the deconstructionists. They claimed to have demonstrated that all idealist philosophy is humbug; they can be deconstructed.

Came a new way of doing science: science with a dogma. All is matter. The proponents call this scientific materialism. They are claiming that this reformulation of science—all phenomena are material phenomena originating from interactions of material objects moving in space and time—makes the philosophy of material monism scientific.

Unfortunately, the truth is the opposite. This idea—all is matter—has already been experimentally disproved. By many replicated experiments. So why do materialist scientists hold on to the idea? The idea is a dogma; they take it on faith. Just like religions, they run crusades of debunking against data that disproves God's supremacy in creating life in six days. Meanwhile the real evidence is neglected or even sabotaged.

Materialism is a dogma. Accepting it as a starting point of science is to compromise the free spirit of science. Indeed, mainstream science has degenerated to a dogmatic materialist science.

And of course, people of this dogma vow to destroys religion, even spirituality, even the idea of consciousness.

But religions stroke back. To people of religions, values are important; God is important. To them, it was science that lost all credibility by accepting an exclusive dogma that excludes God, spirituality, and human values.

The arts lost credibility, too, except for its entertainment value under the materialist aggression. The artists too did not, could not embrace materialist science.

However, degenerating science this way was the ultimate triumph of the rationalists.

The genesis of materialist science still does not make sense! Is it rational to accept a dogma and destroy the spirit of science as the pursuit of truth forever? Without proof that molecular biology really can explain life and produce life in the laboratory? Without proof that we really can build conscious computers like human beings?

Couple of ideas float around to explain the behavior of the materialist faithful. One idea is that many scientists were buoyed by the triumph of our space program. We have gone to the moon and directly verified that that there is no God sitting there like Christianity has been postulating—heaven is out there in outer space. This idea is more like a joke.

One other idea makes more sense. Scientists are a horny bunch. Their anger at religionists is guttural because religions preach morality negating sexual pleasure. When reliable birth control pills were discovered around 1960, that was the turning point. Destroy religion! Let's live a sexually permissive society.

Anyway, this in brief is the story of how we have two worldviews, the denigration of moral values, the devaluing of truth, and the confusion about consciousness and spirituality. Materialists and religionists locked in a power struggle are overseeing weird attacks on democracy, people are experiencing the return of elitism of feudalistic proportion either as the original feudalistic dictatorship or as a new shared dictatorship of meritocrats, our planet is warming up rapidly, some religious people are responding to what they see as the Satans of immorality with terrorism, but all that the materialists/religionists—the rationalists of both sides do is wonder why. Why can't the other side be rational!

This is the problem of throwing the baby out with the dirty bathwater. Allowing the crusade against religion become a crusade against spirituality—something innate in us is perhaps the greatest mistake a segment of humanity has ever made. If you declare rationality is everything disavowing the nonrational, you invite the irrational. If you insist that truth is relative, you get Fox news. In the USA, this is the reason for the meteoric rise of conspiracy theories and media of misinformation.

And then there came quantum science and the promise of worldview integration. Perhaps there is hope for civilization!

More on Quantum Science: Let's Go Deeper

Physics' specialness is undisputed because the movement of objects (where the object will be when) can be predicted by a mathematical equation—the most accurate logic known to us. Newton found the equation of motion of objects—particles—of the macroscopic world and gave us the philosophy of determinism; given where the object is initially with what velocity and knowing all the appropriate interactions the object is under, we can calculate, using Newton's equation, all the future movement.

But the mathematical equation for the movement of quantum objects of the submicroscopic material world predicted that quantum objects can be in many different places at the same time in a way that we call wave behavior. Waves, you may recall from your experience of a water wave move as advancing crests that embrace a whole bunch of places. This is not how Newtonian objects behave! To build a worldview based on Newton's physics as materialists have done is wrong.

The problem is, when we observe the quantum objects, we always find the objects localized at one place except that we cannot determine from quantum physics which place of the many predictable possible places the object will manifest. So, first conclusion: indeterminacy enters quantum physics. This is quantified in the famous uncertainty principle.

This is interesting but easily tractable. We can calculate the probability of where the object will be at a given time. So long as we are dealing with a large number of objects, the probability calculus is highly reliable. Indeed, quantum physics is the most successful predictive theory of physics.

The deeper problem is this: for a single quantum object, what determines where the wave will collapse into a localized particle? A cavalier answer sounds ok: give up causality—the idea that for every effect there must be a cause. But really? Ever since the great philosopher David Hume argued so, the causality principle has been a crown jewel of rationalism. Causal explanation is at the heart of scientific rationalism.

The physicist John von Neuman found another answer: the cause must be nonmaterial. It is true that no material interaction can ever collapse a many-facetted quantum wave to a one-facetted particle, but who says that the causal interaction has to be material? Von Neumann argued that it is the observer's consciousness—nonmaterial to be sure—that does the collapsing. Observer's consciousness chooses the one position out of many available possible positions of the object.

Paradoxes were raised and much controversy. When the dust settled, read Amit's book *The Self-Aware Universe*, the conclusion is definitive and ready with two verifiable predictions (see chapter 2 for details). The consciousness that chooses is everybody's consciousness, that oneness consciousness of spiritual wisdom traditions.

So, at the end of the day, there is light, the end of the tunnel of worldview polarization has arrived. The two verifiable predictions are:

1. Two observers, when correlated by a common intention, indeed become One consciousness, that is they can communicate without exchanging signals. What is special about this prediction is that it is not just about mental telepathy. Since the observer consciousness manifests in his brain, the novel prediction is a brain-to-brain communication without signals.

2. Consciousness manifests in every human observer in two distinctly separate modalities: the familiar ego and one that we call quantum self is the same one that spiritual wisdom traditions talk about.

Both predictions have been experimentally verified, see chapter 2. These predictions and verification are devastating to the philosophy of rationalism/materialism; they put the nail in their coffin so to speak. The philosophical ism that triumphs is the monistic idealism of the spiritual traditions.

Behold! there is no real loser here; everyone ultimately wins because the worldview polarization is resolved and unity of worldview, one truth, has prevailed. So, civilization can progress again.

You maybe wondering, what is so special about the brain that consciousness can manifest there but not in a rock? The answer is crucial.

Reductionist philosophy holds for matter: bulk matter can be broken down to smaller bits: to molecules, to atoms, to elementary quantum objects. And vice versa. Elementary objects make atoms, make molecules, make bulk matter.

This step-by-step continuous way of making conglomerates is called simple hierarchy. The bottom level is the causal level: ultimately, it is all interactions of elementary objects. Biologists talk about genes causing all the miracles of life, but that is *as if talk*. Genes' causal power originates from the interaction of the elementary objects. In other words, cause runs one way, from elementary objects to atoms to molecules to bulk matter, never the other way around. The as if interaction between genes can never affect the dynamics of elementary objects. Simple hierarchy of upward causation.

Rocks are like that—simple hierarchy. They also become virtually Newtonian; that is, as the mass increase, the objects of conglomerates of matter become progressively Newtonian. In a small quantum composite object, the different parts dance like a chorus line, in perfect coherence of phase. As the composite gets bigger, the coherence usually breaks down giving way to decoherence between component parts, like rock and roll dancers.

The difference between a rock and a brain is twofold: 1) the macroscopic, brain retains the quantum modality in some of its actions. 2) When all things are said and done, the brain has two macro components that have a circular causal relationship with each other: each affect the other, affect back, affect it other, ad infinitum. This is called tangled hierarchy.

An example will clarify: consider the liar's sentence—*I am a liar*. You must not miss the tangle, it is a bit subtle. If I am a liar, I am telling the truth; if I am telling the truth, I am a liar. Each time we reach one end of the sentence, there is contradiction that forces us go back and forth.

What happens? If you go in there, you can get caught, if you allow yourself to. It has given you an appearance of a self, separate from everything else.

We know the brain has a perception apparatus, and a memory apparatus. But there is no perception without memory; likewise, there is no memory making without perception—tangled hierarchy. So, as consciousness enters the brain to use its perception apparatus to see, it has to go the memory apparatus to complete the job, only to find that there is no memory without perception, and so on, back and forth.

But you can argue. I can always get out of the sentence's trap, whenever I want to. So can consciousness. What happens? Oneness again, no subject-object split. A state that modern psychologists Sigmund Freud and Carl Jung have discovered as our unconscious.

You can still argue. I can enter the liar's sentence and escape it at will. But not so in the case of the brain. Behold! This is why brain has two selves. When you realize the quantum self, you will find that you can see through the illusion of separateness any time, just as in the case of the liar's sentence. In the ego consciousness, we have forgotten that ability. This forgetfulness of the ego is the creation of further brain processing giving rise to conditioning and me-centeredness. Buddhists intuited this millennia ago and formulated it as the principle of *dependent co-arising* of the ego and the world of objects. We in our ego do not see objects in suchness but always tainted by how we cognize the objects.

Now you see why in spiritual wisdom traditions, people strive so hard to achieve self-realization! The ultimate experiential proof of reality being the way prescribed by monistic idealism is self-realization—that there is no-self. There is only Oneness.

Now that quantum physics has given us experimental ways of ascertaining the validity of monistic idealism, perhaps that self-realization imperative will relax a bit.

Tangled hierarchy! You know who discovered that? It was a mathematician, Kurt Gödel. Mathematics, another crown jewel of rationalism is not fully rational after all, if it has to accommodate tangled hierarchy!

For farther details on how the brain works to manifest quantum behavior, read our book *The Quantum Brain*. For exploring how to rewire and optimize the brain, read our book *The Awakening of Intelligence*.

The Human Consciousness is Relevant

Ultimately, it is the battle of relevance. The reason we study philosophy—being and becoming—is to find the meaning and purpose of human life, how should we live?

Dualistic philosophies are popular because we like the idea of improving our lives. Although it involves at least some attempt on our part to be good, religions always guarantee easy redemption; somebody else is going to take care of us!

Materialism is even better. We don't have to do anything because we can't. Whatever we experience happens to us, we are just experiencing it as bystanders—epiphenomena of the material brain and body. Eat, drink, and survive! Oh, have pleasure too. Pleasure helps survival. If your activities harm somebody, so be it. It is the mechanism. Not your fault. Not your responsibility.

Can reality be like that? Right now, people are torn between these two alternatives, would you believe? Believe it.

Spiritual wisdom traditions have discovered the truth about ourselves millennia ago. It gives us huge potentiality to fulfill but also demands huge commitment and responsibility to get there. So by and large, we ignored the wisdom.

Now science, the newest epistemology on the block, is giving us real hope to live up to our potentialities. Quantum science is backing up the spiritual ontology and epistemology with some highly needed clarifications. Neither the religionist nor materialist picture are totally wrong, but they are incomplete. Also, the means to get to the conclusion were often wrong under both worldviews. When the data is interpreted with the correct latest quantum science, a highly optimistic picture of us emerges. In brief, the new conclusion is we are relevant. Our experiences count. In what way, you ask. Let's count the ways:

1. Yes, we are burdened with a base-level highly confusing human condition that borders on mental sickness; but this

is not because we are material and mechanical but because in the early days of our evolution consciousness has been preoccupied with the base-level need of survival.

2. However, beginning with domesticated mammals, our pets, gradually, our higher needs are coming to the fore.

3. Evolution is purposive, both the evolution of the nonliving universe (this is revealed in the so-called anthropic principle—the universe evolves so as to evolve life) and the evolution of life (the evidence for this is that evolution of life is one way from simple to complex and the humans are its pinnacle) as conscious evolution must be.

4. The purpose of evolution is to express the creative needs of consciousness to see itself in manifestation. Right now, we are on the verge of making a transition to a new era where our purpose will be focused on the exploration and embodiment of the archetypes.

5. We have free will even at the ego level albeit it is only a small fraction of the freedom of choice we have if we can access our quantum self. With our free will, we can say "no" to conditioning, overcome obstacles, and make our intentions come true if we align our intention with the purposive movement of consciousness.

6. Our infinite potentialities are available via our subtle experiences, feeling, thinking, and intuiting. These experiences are no less important than the material world. Learning to value our inner experience is how we become relevant, how we bring meaning, satisfaction, and happiness in our lives.

7. Exploring the archetypes: creatively is how we change from our current transactional human condition to a transformational one.

8. Death is an illusion; we never die except that our story line changes from one life to another to make room for us to live a new story. What reincarnates is the nonlocal memory we create—our habit patterns and character traits and the associated memories.

9. There is a bright future of a more conscious, more loving, more just, more whole, happy and intelligent human society ahead of us. We are participators in creating that future.

If these words inspire you are ready to participate in the great adventure of consciousness, of God, of soul-making.

CHAPTER 2

What is Consciousness?

An Upanishadic story first. Two students, one a materialist, one with open mind asks a teacher about self-knowledge, who am I? The teacher says that one's own image that one sees when looking into the eye of another, or into the water of a lake, or into a mirror, is the self. What the teacher meant is that if you look into a mirror you see yourself clearly as the experiencer of seeing your image. Both students misunderstand; they identify their self as their body which is what they see in the mirror.

However, the post materialist has doubts; he goes back to the teacher and the teacher gives meditative practices to further look into the question.

Incidentally, today psychologists give the test to one-year old babies except that they put a huge red mark on the chest of the baby; whereas a monkey would go and look behind the mirror, the baby is not fooled. Indeed, all babies at about that age can see that the image is of their own body. In this way, they have learned to distinguish themselves and the object.

But ask the baby the subtler question, who are you? She will indicate her body. It is quite amazing though that people of a materialist mind set can never get over this initial confusion, I am an object, my body.

What indeed is consciousness? you don't think about consciousness much do you? it's not our everyday type of thing to discuss what is consciousness. If somebody raises that question anyway, you would probably say let's leave it to the philosopher. I don't want to deal with it. What makes you think that philosophers can do better?

But seriously, it is an important question because everything you know about the world that comes to you through your consciousness, right? The rock does not know, no conscious awareness for them. Animals, latest consensus says that they too have consciousness, what does that mean? How is animal consciousness different from ours? And first of all, how do you know that you are conscious? what difference does it make, if any, to the affairs of the world? These are important questions, right?

In fact, you do think about consciousness a little bit when you deal with the idea of free will or just freedom? You know we live in a society and society puts restrictions on us. Do we have freedom to choose whatever we want and we can do that or are we constrained already? Maybe, society is just codifying some of the constraints.

Some people say we are machines, material machines, mechanistic; we don't have free will. Have you thought about that? What is the consequence of your being just a mechanical person? People wonder because if it is true that we are mechanical machines then questions arise about whether we should be punished for misdeeds that we do. Because after all we are machines; we are not doing misdeeds because of our free will. Intricate question. How far or how seriously should we take this proposed zombiehood of ourselves? Because, in actuality what happens is that your actions very much determine your relationship to the world. If not to the world at large so much, but the small world defined by your family, by your intimate relationships, right?

Nobody can deny also that there is this grammatical question as well. We do use the first person I to describe our experiences. When you eat you say I eat; you don't say eating is happening to me. I eat. Hopi Indians (the native American version of Indians) are supposed to have eliminated the word I; they will say eating is happening, but that doesn't say who is eating so you must say eating is happening in this particular mouth in this particular body-brain combination that defines an object "me," right? Instead you say I am eating; I am experiencing my experience. Does that have consequences? it sure does; it keeps your body healthy that part is conditioned, objective. But if you're eating, that affects other things too! your relationship to yourself, for example. You are happy. Something else! have you noticed that after eating you become a little bit more generous? what does that mean? in your generosity you might be a kindler or gentler person for a while. You're happy with yourself and therefore if your children ask you for something or your spouse, then you are more likely to go along with that. It means that your consciousness has become more inclusive in those moments. So you see it is an important question what is this consciousness, this experiencer "I" that can expand to include somebody else? It's not just "I," but the I can become "we." How does it become we? What is the relationship between I and we? So, it's not such a simple proposition after all—saying that the world is mechanical, it's all objects; there is no such thing as a subject; or I is just an ornamental grammatical description that we give to an experience that is happening, that there really is no body experiencing meaningfully.

So again, the way of thinking about consciousness as ornamental epiphenomenon just does not match your experience, does it? Because that kind of definition would restrict yourself to the machine, to only the object that you call me. This capacity that you have, to extend yourself to include another, to become an we, that is something very special and that suggests that the subject-consciousness "I" is something very special isn't it?

There is this opposing view to the machine view because you know many people not only experience expansion of consciousness just simply while doing things like a pleasureful activity or a positive activity, they experience consciousness in that way most of the time anyhow. You don't believe it? Believe it because some of the most revered people in our history belong to this category; people that you have heard of, almost everyone has heard of; Jesus, Buddha, Krishna, Muhammad, Rabbi Hillel from the Jewish tradition, Guru Nanak from the Sikh tradition, Rumi from the Sufi tradition; the list is very big. Every spiritual tradition has people who have reported unambiguously that they experience their consciousness in that expanded form all the time or most of the time at least. Is that a misrepresentation, an exaggeration? just a way of making themselves more interesting, make themselves more intriguing to their disciples? But you know that would be a very myopic view to take. Our religions are all based on the teachings of these wise people and the spiritual wisdom tradition that they established and taught are still alive in the esoteric branch of all these religions.

In the popular front, of course religions have become very ambiguous about the possibility of such expansion of *our* consciousness. They have a concept called God and usually their overt statement about God is not about expanded consciousness or oneness; instead, as defined, God seems to be a person like us, only super-duper. In popular depiction in the West dominated by white people, because Jesus, the founder of Christianity, referred to God as "my father," the tendency is to depict God as an old white male with a beard like Santa Claus except that He is much more powerful to denote which they put some evocative adjectives. Those adjectives describe God as almighty, omniscient, omnipotent, omnipresent, like that. Behold! Isn't omnipresent synonymous with everywhere—expanded consciousness or One interconnectedness of everything? So, if we think about it, we get it. But if we don't, then elitist people are going to sell us a whole bunch of crap about these ideas about ourselves and distort them to short-change us.

For example, something that we all want—free will. Can they take away our free will? You bet. Materialists do it by the clever proposition that we are machines; religions do it by proposing that it is the omnipotent God's downward causation that make everything happen; you have no role to play.

This is what dictators do in various countries. There are dictators because people today are very confused if they have free will or not; they are looking for a leader to tell them what to do, to take them to the promised land; they have given up the idea that they have personal power to choose and make a better world for themselves and others.

A MAGA follower dies of corona virus and goes to heaven. He was rushing through the gate, but St. Peter stopped him, "In heaven, everyone has to wait their turn." And just then a figure in a MAGA hat goes running through the gate. Out MAGA guy was furious, "There! you let him in." St. Peter Chuckled. "Oh, that's God. Sometimes he likes to run around in a MAGA hat. He says it empowers him."

Politics and Religions have always been run by elitists defining aristocracy and religious oligarchy respectively; and now scientists of the materialist ilk have joined the elitist club, defining an elitism called meritocracy. Knowledge is power.

How did this happen? that too is a good question. How can we avoid being manipulated without knowing who we are, if we have freedom, okay? So, these questions of consciousness are serious, are important; they do govern the way we are going to build our society and have built our society.

Look at the democratic constitutions. Almost everywhere, the most explicit statement is the one in the American constitution where we find the statement, all men are created equal. Okay, it would be better if it said, all people are created equal, but this was at a time when thinking and governing just was men's territory all over the world.

Let's examine the idea all people are created equal. What does that mean? where does the equality come from? why is that important? It is important of course because everybody gets equal voice in choosing their representation, a process that we call voting. Everybody has the same vote in a democracy, and this is the best institution so far that we have created compared to other institutions like monarchy or dictatorship that still are around.

There is political crisis right now in many countries including the USA, but democracies are surviving. So, we are making progress; we are acknowledging that there is equality among people but where does that truth come from? Certainly, people are not equal in physical strength nor in mental strength nor in opportunities and power and all that! there is social gradation everywhere; economic difference everywhere in all societies between different statures of people and still the idea that all people are created equal have survived. Amazing because the idea is an acknowledgement that maybe people are based fundamentally on a oneness, that expanded consciousness that we talked about.

Where did this concept of oneness of consciousness originate? it originated in India codified in books that today we call Upanishad or Vedanta. Some one hundred and eight Upanishads exist and each of these book's authors did not exactly write it down; the wisdom was passed on in spoken words for a very long time, eventually codified only when writing and paper became fashionable. What is certain is that the researchers and authors of these books endorsed the truth all is *Brahman*, Brahman is the Sanskrit word for consciousness. Declared one of the authors of the Upanishad, *Sarvam khalyidam Brahman.* Those are the Sanskrit words meaning all is Brahman; Brahman is Oneness.

Later another great Indian philosopher Shankara elaborated Oneness further as "one without a second." Elsewhere, the Upanishads go a bit further and amazingly gives a definition which still can raise our envy with the kind of clarity that all covet. The definition has Brahman having three qualities: *Sat-Chit-Ananda*, Sanskrit words once again. *Sat* meaning existence; *Chit* meaning subject-object awareness, and *Ananda* meaning joy, but the etymology of this word Ananda requires some elaboration. Etymologically, Ananda comes from another Sanskrit word called *Ananta*. *Anta* is boundary; Ananta is no boundary, so what is the idea of Ananda? as you get rid of boundaries of your consciousness, expand your consciousness as you increase the boundary of your consciousness, there will be more and more joy.

So, the definition is telling us the nature of consciousness; it is laying it out right there. It agrees with your experience. Your ability of inclusion is much reduced in a day when you are grumpy, and you have some health issues and things are not working out as you expected. On such days, you'll find that indeed you are very me oriented, very selfishness oriented, you are not generous. Compare that with another day where you are relaxed, you are on vacation, you are having fun with your family, and you will see that your mind is full of generous thoughts, "oh I love my family! oh how beautiful my daughter is, oh how great my wife is, my husband is, my son is. You know even other people would look beautiful to you, you would would more easily smile and say, "hi good morning." When that happens, you are including many more people in your consciousness, it has expanded, okay?

Ananda—increasing levels of expansion of consciousness— defines our spiritual journey; spirit is that ultimate expansion that oneness is. This is the idea: the existence, subject object awareness and then this joy of expansion that today we identify as happiness, define consciousness.

How is happiness different from pleasure? Pleasure can be called a molecular phenomenon; there are molecules that are emitted in the brain called dopamine that gives us the sense of pleasure. Eventually pleasure can end up in happiness too, so, people often use the words synonymously, but the experiences are different. Pleasure is more intense. But after a pleasureful experience, we usually end up happy so there is an ambiguity there. But remove that ambiguity; it would help our discussion in the future. Pleasure denotes that intense experience that comes from molecules in the brain, molecules in the body hormones, but then there is this aftermath of pleasure; then there's just a general glow, an expansion of consciousness felt as generosity and that's happiness. For some people, pleasure is it; after that they still retain grumpy consciousness. So indeed, it is better to distinguish between pleasure and happiness although we notice that there is a connection: pleasure can lead to happiness.

Okay, so consciousness is defined as existence, subject object awareness, and then increasing happiness as we progress in the journey of consciousness. This is the idea that gave us the spiritual wisdom traditions. The wise lives in more expanded consciousness; only the ignorant lives in the contracted consciousness which is often compared with darkness, okay?

Let's go to the opposite end. It is more or less true that most people live in very contracted consciousness especially today. It is not very clear why that happened; people have some theories for it, and I will tell you our understanding of it during the course of this book, but the fact is that yes, we certainly have developed a sociocultural situation where we live in constricted consciousness more and more. This shows up in people being unhappy.

There's a story about Mother Teresa when she came to America from Kolkata where she lived. Kolkata as you know is full of destitute, lots of beggars, and was especially so in Mother Teresa's time. She was not St. Teresa then of course. She died as Mother Teresa. When in America, she was asked, you are needed in Kolkata, why did you come to New York? we don't need you here, we are an affluent society, no destitute here. You know what mother Teresa said to that? I see more unhappy people here. This surprised a lot of people but this is true. In spite of all the affluence in the rich countries of the world and America certainly is one of the richest, in spite of all the affluence, people in these countries live in very constricted consciousness. Well, one reason is the brain. You know we experience consciousness through our brain and this is why many scientists think that brain is it, brain gives us consciousness. They objectify consciousness. The brain is very negative, five times more negative than positive. So there you go! Negativity, negative emotions that we experience from the brain lead to constricted consciousness that identifies only with this individual body/brain.

From this constricted consciousness, it is difficult to see that there is that subject-object polarity of awareness; it's difficult to see that there is an "I" lurking behind "me." This is why some very serious scientists, serious thinkers concluded that maybe we have a straightforward science—all is matter. Then we can simplify things and straighten out all these questions of reality with this one assumption. Maybe religions are humbug, we have difficulty finding scientific basis for them because there is no scientific basis. Consciousness, spirituality, moral values, all are just people's imagination. Maybe what people talk about when they talk about expansion of consciousness is just something of a fantasy!

Do we really need to assume that there are non-material entities in the world such as consciousness? this is the debate. Is the I—the subject of our experience—a self that is separate from the objects that it experiences? should we continue to use that I as if there is an experiencer and take the experiencer/self seriously? Instead, why not just assume that I is also an object because I can certainly look at my "I" and then "I" is just "me—an object.

But wait! Experimental scientists are a very honest bunch in many ways, and they pointed out correctly that the individual experience is always a little bit individual that cannot be eliminated, that cannot be described objectively as a brain phenomenon, there is a little bit of personal experience—a subjective felt quality. It especially shows up in the way we see color. You know physicists did a lot of study of colour vision—how we see color. There are lots of theories; we don't need to get into the details of theory, but it is a fact that we see color in a very personal way; color without vision is just impossible to make a science of. We have to make a science of color and vision that means including the personal quality.

So, in the least there is this subjective quality of the experience, quale (plural qualia) is the technical word—subjective quality that even materialists struggle with. The I, the self, cannot be completely thrown away.

The Upanishads of course say that I—the subject and the object(s) of experience, they really are coming from a Oneness, spontaneously arising from the Oneness. Subsequent conditioning obscures the oneness. Buddhism makes it a clearer declaration: *patitia samutpada*—Sanskrit for dependent co-arising. The ego experience and object arise together from that one reality which in Buddhism is actually identified as *Nothingness* but that's another discussion, okay?

What is Transcendence?

What is consciousness? is a very confusing question because there are so many claims, and often they go in different ways. Above, we have given you a glimpse at some of the controversy that dominate the question today. There are thinkers that subscribe to the idea that everything is matter. Most scientists are among these thinkers. And because their view of the world is that it is all made up of matter, there is nothing else, the view of consciousness from that angle can only be that consciousness is an object. Whereas if you think about it, then certainly, one can argue that no, there is also a subject pole, an experiencing self, of every experience that we have.

Conscious experience is the way that we know that we are conscious. Right? Most people know that they're conscious, and others are too. Because they have conscious experience, they assume others must be having such experiences too. I should say all people here except that for some, other things such as me-centeredness affect their conscious experience so much that they can only think of themselves; others are object for them to manipulate. These people are close to solipsistic, behaving as if only they are conscious.

Consciousness can only have experience when consciousness splits into self and other; one part experiencing the other. It is impossible to *know* Oneness and its potentialities staying in Oneness. This is why the play of manifest experience.

Does the idea of an experiencer make sense now? There is an object of experience and there is an experiencer, a subject that we in fact refer to as I that is looking at the object. But when it comes to understanding this, you can easily get confused by asking, but can I look at me looking at the object? Yes, you can and you do. We all do. And in that way, your "I" can become an object too because it can be looked at by you. You call that your "me." Now, this would be no problem if you could discern between I and me, the confusion is created because what's the difference between I and me is not easy to answer. Think about it very hard. What's the difference? Because everything that you do you say you think you experience as I, add to your me and again, What's the difference between the two?

Because in your ordinary station of consciousness that you live, you cannot really differentiate between your I and your me, and since this is your most common experience, you just will concede that consciousness could be an object when a scientist or a philosopher says so.

You are being too hasty. There is one more thing. And this even the materialist philosopher or scientist agrees is a touchy issue. You will agree that your experience is personal when you are looking at a rose, right? Your experience may have similarities with other people's experience of the rose, if they are living in the same culture, same society, same community as you, but still there will always be a personal element in every bit of your experience. This is one thing. Essential subjectivity. There's a subjective quality of experience.

And there is a second thing also; you would reluctantly agree that there are occasions during the day when your consciousness that experiences the I seems to have difference from the me of ordinary experience. What kind of difference? The "I" seems to be able to expand to include others and if those others are your intimate relationships, you often describe these others fondly as "I love them." In those moments, when you experience loving them It really is different, right? Not only you think that I care for this person, but you also feel something in the body, often involving the heart. Women more than men, to be sure, report feeling energy in the heart; this is one reason why women are more into relationships. But even for men too, undeniably there are occasions when that inclusive ability of the "I" shows up. I can make a we on those occasions; this the "me" can never do, right?

So, these are the two ways that should keep bothering every reasonable thinking person thinking about consciousness, okay? So, what is consciousness to you now? Not an object anymore, is it?

On top of it, there are a few people even today, who declare squarely, and you can trust them because you know, there are ways to tell that whether people are lying or not, that they more or less regularly experience their "I"-consciousness as expandable and claim that their "I" is quite different than their "me." In fact, there is a difference in behaviour that you can discern in these people; often they're able to behave in what we call an unselfish way that you mostly can't.

To 85% of today's people, this is not much of a news. So what? But if you belong to the 15% who care about questions of reality, then you know that your behavior is mostly transactional, rarely inclusive of the other, and even when it is inclusive of the other, you often find your inclusive behavior may not be unmixed un-selfishness, you hesitate to act on your momentary feeling of inclu-sivity. Whereas these people seem to be acting unselfishly without making much effort, or without that hesitation. How do they do it?

Here as well the materialists have an answer. Materialists say what we do when we talk with people is exchange signal going through space taking a little time; this is called locality. The signal goes from me to you, and you respond, and we have a conversa-tion that's local communication and that communication can be done nicely. If it is done which some good usage of language, using words like love and stuff, then you and I establish a friendship. You can then make believe that your dealings with this other person—me—is unselfish. This make belief is a capacity of our brain. Brain cannot differentiate between such make-belief imagination, and the real thing. All relationships like you have with your intimate relationships, according to materialists are make-belief unself-ish. Your experience of expansion notwithstanding, these are just examples of such socially convenient relationships laced with a lot of imagination. So, these arguments go on and on and on and on between the idealists and the materialists. People who give value to ideas of consciousness such as love since they have experienced the expansion of consciousness and inclusivity and people who live so much in contracted consciousness that they feel compelled to rationalize away any love-talk as pretend behavior.

What is basic here is this question: Is your consciousness really expandable? is there "I" beyond "me" that can expand and include another? Can you act in a non-transactional way in an unselfish way at a genuine aspect of who you are, of your consciousness?

And surprisingly, science has moved on beyond the materialist science and is giving us affirmative answers to these questions: there is something to this expanded nature of consciousness. After all, it is just not spiritual traditions telling us this, only spiritual people reporting their experience this way, it is just not imagination of some people. It is now experimentally verified and have been for years now. What do you say to that? The experimental data was predicted by the new physics, quantum physics. There has been a breakthrough, which is we want to communicate here.

What we have gone through so far is just a preamble. The preamble is important to show the problem. But once you understand the problem, then we can tell you with confidence that our new physics has proven beyond any reasonable doubt that there is something to expandability of consciousness. There's a new concept on the block: communication is not only local, always using a signal that goes through space and time from one to another. There are also communications that quantum physics permits that are *instantaneous using no signal from one to the other*. We can only communicate with ourselves without any signal. So, this communication from one to the other without signal can mean only one thing. This otherness of the other all of a sudden has disappeared. The two people who are communicating in this way are using nonlocality. When there is no local signal, the communication is denoted by the word *nonlocality*. It's a new word, it's a quantum word. The two people via this ability of non-local communication have become one, their consciousness has expanded, according to experimental data. So, there is no way that we can refute it. The only question is: what allows this nonlocality? Are we understanding it properly? Is there is a theory?

Quantum physics gives us a theory, the experiment that confirmed nonlocality was predicted first among objects, quantum objects, and then between people.

Now do you see why the question where your consciousness ends needs to be asked again and again and why if you have any belief that you—your consciousness—ends where your body ends and you end where your brain ends, give up that belief, give that all up, just give it up. It is just not the case. Your consciousness can expand beyond your brain to engulf another person; in those moments you become one with the person that you are communicating with without using any signal; you have established a non-local expansion.

Yes, you have this capacity of nonlocal expansion of consciousness. In those moments, you can be very generous. In those moments you can give and there will be no reservation, no constraint, or conditions. Reservation will come later, of course, your consciousness will judge you later. And then you will say, what happened to me? Why was I so vulnerable? You might even think that I should not do what I did, because, you know, many people are very shy about establishing such intimate relationship because they've, indeed, found themselves in such generous situation, becoming generous momentarily, and then later on regretting, and then they don't want to do it again. Some people even develop an interior castle, kind of; you can only come a bit close to me and be intimate, but beyond that is my personal space, I will retain my me-ness in my castle, that is my right. And you know, it is their right. That too is part of the equation of consciousness, we have the choice; this too is not part of the materialist approach to consciousness, because there we are just mechanical machine with no choice, but the new view is telling us that yes, there is choice. Even a sociopath has a choice; he can change.

So, in this way, what is happening in science is very beautiful. Quantum physics is coming along, having settled this basic dispute between the two polarized views idealism and materialism, with both theory and experiment.

Mind you, I have not given you any details. We'll wait for that in later sections. But with this basic dispute settled, we now can move ahead. And settle the rest of the questions raised above. The rest of the questions are also very serious. Rest of the questions about the worldview, about our society, about our culture. You know, right now, there is a huge polarization; huge polarization between people who think all-this-is-matter kind of way and people who think human being are connected to higher power.

Materialists are followers of materialist science. People who think that no, things are not this way, only material, they are followers of some kind of religions—offshoots of idealist thinking. Both sides have something right on their side and something wrong on their side. But both sides of dogmatic thinking will always insist their dogma is correct. The scientific dogma —everything is matter—is under challenge since 1982 when quantum nonlocality was experimentally verified since material interactions can never simulate nonlocality. But the dogma retains its grip over its followers, who resort to wonderful sophistry to rationalize nonlocality in human experience away. So steadfast is their faith in materialism.

And then these same scientists criticize Christians for their steadfast faith in the Book of Genesis in spite of the fossil data.

Let me repeat. Quantum physics already has shown that this philosophy—everything is matter—is false. Material interaction can never be non-local; everything is local, in dealings with matter. Where does nonlocality come from? If there's only matter, it could not arise. As simple as that. Very interesting.

Let me repeat. Religions also have dogmas. So certainly, nobody in the right mind should ask you to accept one dogma for another. Such a trade will not be very profitable. What is dogmatic about religion? Let's go on and examine that some more.

Religions came from the teachings of great masters who are called mystics today, figures like Jesus, Buddha, Krishna, Mohammad, and others—Lao Tzu in China, Socrates and Plato in Greece, Muhammad and the sages of Kabbalah in the Middle East. What these people taught one and everyone is about the oneness of everyone, oneness of everything. All is consciousness, declares the Vedanta, enunciated some 7000 years ago in India. Jesus discovered and repeated the same truth two thousand years ago. Buddha, Socrates, and Lao Tzu, a little bit before Jesus, Mohammad is more recent. The written version of the Kabbalah is more recent too, but the ideas in it might have much older origin.

But again, not only these ancient people have said such a thing, people today can experience this oneness and when they do, they declare the same thing. Mystics abound today. Krishnamurthy, Anandamayi Ma were alive as recently as in the nineteen eighties. The fourteenth Dalai Lama is alive today.

It is even true of scientists. Not all scientists buy materialism. For example, the psychologist Abraham Maslow. He has said that there is consciousness beyond ego and in that consciousness, you become one with everything, he calls it peak experience. So, such ideas have come from psychology, too. Not just quantum physics, where Amit discovered that the primacy of consciousness is the solution of quantum physics' puzzles and paradoxes from direct creative experience.

You will ask, where is this oneness located? this is really the problem, where is oneness, we certainly do not see it here in manifest reality? Where is it located then? And this problem really has not been solved by the spiritual wisdom traditions in a very understandable or convincing way.

Well, they have tried, the mystics, and later even religionists. For example, the word transcendent and imminent are often used to differentiate between the two types of statement; Oneness is a transcendent oneness we are told and the observation "I am separate from you" is an immanent separateness. But where is transcendent nobody would say so, very clearly in the past, nobody knew, no adequate concept was there that would make sense to the rational mind. This idea of transcendence is *super-rational, not irrational.*

But again, the rational mind mistakes the super-rational as irrational; so, religionists eventually came around and they simplified the issue. That is, when the religious dogmas were create that later scientific research refuted via experimental data.

Let's take the case of Christianity. The Christian thinkers looked at earlier Greek thinking for guidance as to how to talk about transcendence. Greeks also had to face the same problem earlier; there were Socrates and Plato, their teachings and of those subsequent to Plato such as Plotinus in Europe were the same; they were talking about this interconnected oneness as transcendent. Where is it, this transcendent? What is transcendence? no answer. Same problem everywhere. But what happened in Greece after Plato is that transcendence thinking was rationalized in Aristotle's philosophy. Aristotle being a realist said Okay, so this transcendent Oneness must be in a domain within space and time where we cannot go. We are here on earth. And then there is this vast Outer Space; could the latter be the domain of oneness? And later when Christian rationality had already simplified the Oneness and identified it with another popular idea from Judaism of a monotheistic God, Aristotle's picture was a perfect fit. Jesus himself had called the Oneness "my father." So, who can blame Christians even today thinking of God an old white bearded male sitting in a throne in outer space which is heaven?

If we say that this transcendent world is outer space, this is where science beginning with Galileo, started to raise doubts. Newton developed the science further to prove conclusively that no separate laws of outer space objects—the sun, the moon, the stars; the physics laws they obey are exactly the same as the laws for earthly objects. Newton's laws of motion and law of gravity are the same everywhere. This left no doubt that there is no difference between the Earth and the outer space—they are all part of the immanent universe with one set of laws. Difference is just the distance—vast distances. But otherwise, it's the same space and time everywhere in the universe.

Then what? Christian religion had made a choice. They had converted oneness, into a monotheistic God that integrated all the different earlier concepts of higher power—little gods, now we recognize them as archetypes, a concept created by Plato. They even found heaven, the abode of God. All those concepts—by now dogmas—were now in jeopardy.

A God with a big G, not little g gods, not the archetypes, but the totality of all the archetypes plus everything in existence is God's governing territory. And that's God. Is there anything in the religious depiction that shows that the Christian thinker were not exactly creating the popular picture that Galileo and Newton refuted. There is. There are adjectives that we must use to describe God, we were told. Oneness was replaced by the idea, God is omnipresent. Now look at the concept, omnipresent. How do you think of omnipresence? In popular thinking, God is a person, just like the human person. After all, we can only think of God in our own image, very difficult to think any other way, it will always be about images that we can imagine. I cannot easily think of things that are not in my vocabulary, not in my wildest imagination. So, this super-duper human person that we think is God, capable of this omnipresence, capable of omniscience, of knowing everything, capable of omnipotence, creates and controls everything, but we still think this is a super-duper human being we are talking about. We think, omnipresent just means that this super-duper being like thing is present everywhere. Have you ever seen a human being capable of being in more than one place at a time? Admit it! Your thinking is too simplistic; God is nothing like a human being! It is a transcendent being, transcending your limited notion as well as transcending space and time. How erroneous is the caricature that this being as an old white male with a beard?

And once this kind of erroneous pictures arise, then other errors also arise. For example, how does this being rule over us? Because this super-duper being also is omnipotent, all powerful almighty, He rules by human behavioral rules of control: reward and punishment! Fear of punishment, mainly. If you don't obey, you go to hell. If you obey you go to heaven, that's the proximity of God, that's a reward. Now, normal Christians just go on taking these words in this way, because it makes sense, the popular version of it. And of course, modern people with even a little science background see the picture as a caricature.

A rabbi dies and goes to heaven of course. The place seems wonderful, but he wants to find his teacher first whose guidance got him here. A token of gratitude. He finds his old rabbi as usual reading a book and taking notes at a desk. Nothing special, same old. Then he notices the bed; a beautiful woman and obviously amorous with desire. He winks at the rabbi, "That must be your heavenly reward, huh rabbi?"

The Rabbi does not look up. Wryly he says, "No son; I am her punishment."

Is there any other version of God that makes scientific sense? Yes, of course, there is. You have to just think adroitly, what is omnipresent? Omnipresence is a presence everywhere, but it does not have to be a person being present everywhere, it is the presence of Oneness everywhere. Still problem, still problem, we don't see the oneness here, in immanence. This is why they called the oneness transcendent with the hope to keep a mystery around it so people do not succumb to any simplistic caricature. Unfortunately, people do, people have.

We have not solved the problem yet. Just calling the oneness so obscure that we cannot see it calls for confusion, not clarity. Just like Jesus said, Kingdom of God is everywhere, but people do not see it. You cannot see it in your ordinary state of consciousness; you have to meditate, go through an exploratory journey to experience. That's what Jesus was trying to tell us. This is what he had to do; if any person seriously asked how, he would say, "follow me." What he implied is, "Meditate. Be creative.

Can we understand transcendent Oneness without following a teacher, without a meditative exploration? Today we can. This is the problem that quantum physics has solved. Today, we can understand Oneness by the following statement. This Oneness is not the case for space and time; this oneness happens in a domain outside of space and time. It is a potential nonlocal connectivity between quantum objects of potentiality. Call this domain of reality outside space and time the domain of potentiality.

The concept of nonlocality between quantum objects of potentiality not only solves the problem of, Is consciousness expandable? but also the question, Where is Oneness? To repeat, Oneness resides in the domain of potentiality outside of space and time.

In fairness, you know who else had discovered such concepts in modern times? I shouldn't say who but a whole class of people, psychoanalysts, beginning with Freud. At the tail end of the nineteenth century, Freud was the first to intuit the idea that there is an unconscious behind our conscious experience. Freud was worried about why people suffered from mental disease and his answer was that the aberrant behavior comes from the unconscious. Freud proposed that psychoanalytic therapy opens up your unconscious, and once the unconscious was revealed to you, people would be miraculously healed. Psychoanalysis worked, although not always. Nevertheless, unconscious became a part of our regular vocabulary.

I'm sure you know what the word unconscious implies: Consciousness exists in two domains. In one domain, the unconscious, we cannot experience; we are unaware, there is no subject object split awareness, there is no one aspect of consciousness, self, experiencing another part, the objects. This unconscious quantum physics identifies as the domain of potentiality. Quantum physics is also generalizing the concept of the unconscious to extend beyond one's personal as in Freud's conceptualization to all potentialities available—past, present, and future.

In the space and time domain, we can experience; in quantum physics that is called domain of actuality. Locality reigns in this domain. We are conscious in this domain that we are conscious; we are also separate in the domain. In the unconscious, we are one although like all else in the domain, this also a potential oneness. Objects need to be correlated or entangled before there can be experience of oneness when manifest.

Now a coherent picture is emerging, I hope about the concept of consciousness, and where is oneness versus why we normally experience separateness. The basic puzzle about consciousness, about oneness, comes from the fact that we don't see ourselves to be one with another in ordinary experience. We don't experience it that way most often. And therefore, doubt enters. If we are all one, why are people separate? Mystics, even some religionists, they're always saying things like we are one, but we cannot experience the oneness easily. And that bothers us. And this is why bulk of us fell for the materialist thinking when scientists offered materialism as a dogma, as if it was verified truth.

Some of us trusted scientists because they were saying something that better explains how we experience consciousness ordinarily, and because they were supposedly following a tradition of solid theory plus experimental verification before one elevates a hypothesis to be a law. Molecular biology had no explanation of life, why living creatures have a self, that enabled them to experience themselves as separate from their environment. Ditto for human consciousness; molecular neuroscience had no explanation of why we experience a self, separate from the objects of our experience, that although ordinarily shrouded, reveals itself in moments of expansion and inclusion of others. Instead, all there was promissory materialism, as the philosopher Karl Popper properly called the materialists' bluff!

Science is so divided today that one compartment has no idea of what is happening in another. As a quantum physicist, I (Amit) took molecular biology for granted, did not realize that this was just a unverified hypothesis being floated.

But no need to despair. Understanding is coming from quantum physics itself now. How we can expand our consciousness, how you can expand your consciousness, is being explained by the new physics; how the oneness, become separateness is also explained by the new advances in quantum physics.

And so, how do you really exist? You exist as your me-centered ego, but that is only the base-level of your existence. That's your general normal consciousness, but you also exist as one super-duper expandable consciousness with infinite potentialities to explore. We will take up this idea further in the next section.

Does Consciousness Create Reality? Do We?

So, the question of consciousness becomes more and more interesting. And new surprises emerge. What you originally may have thought from your experience of mostly me-centered consciousness, modulated only occasionally, in fact, probably quite rarely, with an expansion to include another, is simply half the truth. It now seems that that's not how consciousness begins in us. When we are born, it is very expanded; so much so that it resembles the unconscious unmanifest Oneness. It seems that that oneness manifests in the world of space and time as this highly expandable self. And then gradually becomes constricted, becomes individual becomes separate from the rest of such manifest selves all going through the same conditioning experience. That is the picture.

How did quantum physics get to that picture? This is really an amazing development. You have to know; Newton was the first to put together a science of material objects. And that's the key— material objects; physicists figured out that the world of objects that we experience out there is explainable with Newton's calculus and therefore, must consist of these material objects. Newton was the first to figure out the equations—mathematical equations of movement, calculus, of these objects. He was able to calculate where would an object be and when. If everything about an object's future is known from its present condition, initial conditions, and the laws of the object's motion, which includes the concept of force, then we know everything that there is to know about every object in the universe in principle at all times. The picture that emerged is, by and large, the picture that we have today, with some modifications to be sure. But as far as the basic philosophy is concerned, the world of science is the world of objects. Except that question of consciousness, who are we? which is a major question, is where science of objects is inadequate and must give way to a new science that can handle both subjects and objects and their source—Oneness. And that is the role that fits hand in glove what is emerging from quantum physics.

Even quantum physics started as the science of objects—nobody suspected otherwise; many scientists still cannot come to terms with the idea that quantum physics begins a new expansion of science to include consciousness. Matter in bulk can be divided and divided, ultimately, reaching a stage where they could not be divided any further and still retain the macroscopic essence. These objects are the molecules. Even molecules could be divided in atoms, and we found that even atoms can be divided further into elementary particles—the building "blocks." Quantum Physics grew out of the realization that this realm of elementary particles reveals stuff that Newtonian physics has no explanation for. Scientists had to discover new equations. And those equations were discovered. That's quantum physics. It was discovered by two great physicists, Erwin Schrodinger, and Werner Heisenberg. Other very famous names contributed to the development of quantum physics: Max Planck, Albert Einstein, the most famous modern name in science, Niels Bohr, Max Born, Louis de Broglie.

The concepts that emerged from the quantum equations are amazing. They did not emerge overnight. They emerged in tandem with experiments that were conceived, that were carried out sometimes later, sometimes sooner.

First, quantum physics is about elementary objects of both energy and matter and they both satisfy the quantum equation the solution of which gives us a description that we call wave. Waves spread if left alone, if free to move. They'll spread and become bigger and bigger, be able to be in very different positions all at the same time. That agrees with your experience of waves, doesn't it? Look at a water wave; see how the crest lines are spreading, engulfing many different points of space at the same time. This is how we hear sound all of us together, while listening to a speaker. Whereas a particle makes a trajectory when it moves, being only at one place at a time. How Different a particle is from a wave! And yet the particle picture of the quantum object doesn't go away either; when you try to measure these waves of the quantum objects, they become particles, they localize at one point of space, not smeared over an entire spherical crest, an amazing situation.

What kind of waves are these quantum objects then? If we can never see them as waves when we measure, where are the waves? Moreover, the collapse upon measurement from wave to particle is instantaneous. This means that the waves are not residents of space and time at all; in space and time, nothing can move instantly, nothing can move faster than the speed of light. So says Einstein's relativity theory and experiment after experiment has verified the idea. In order that quantum physics does not contradict another lawful theory, we must postulate that the waves reside outside of space and time.

Quantum objects which in their suchness are waves, *really* reside in a domain outside of space and time. They are waves of possibility; quantum math enables us to calculate the probabilities associated with the possibilities. That is how quantum physics proves its predictive power that gave rise many new technologies such as lasers and computers.

How can we discern something like a domain of reality outside of space and time? How can we measure experimentally the defining quality of such a domain? This question raised its head and the answer quickly emerged: the discerning quality is nonlocality.

Recap: this domain where objects exist at waves of possibility is that domain of potentiality previously introduced. It is between two objects of potentiality that we find nonlocality if the objects are *correlated* into oneness. This world correlation was introduced, also the word entangled was used to describe the state of nonlocality. How do two quantum waves of possibility become correlated or entangled? if the two objects interact. If they become thus correlated or entangled, they can communicate instantly at a distance without exchanging signal. This is that concept of nonlocality, right?

To repeat for the umptieth time, the defining quality of this domain of potentiality is this: any two quantum objects of possibility, if they become correlated by interacting, then they have this oneness because they can communicate instantly irrespective of how far they are from each other when they are revealed in present time, when they are actualized in space and time. This is Oneness-at -a-distance.

And this theory was given in 1935 by no other than Einstein himself with two associates Boris Podolsky and Nathan Rosen. The experiment of verification was done in 1982 by Alain Aspect and collaborators, Dalibar, and Roger. It took that many years and many workings with the original theory before physicists could devise a way that one could experimentally verify.

There it is, now, the idea is verified. There is definitely a domain of potentiality where waves of possibility reside. This contradicts the materialist dogma that there is no domain except space and time.

Any two waves of possibility can be correlated to achieve oneness by simply interacting. So, there is potential oneness in this domain. How do we prove that this Oneness is consciousness?

The first hint came from another great physicist/mathematician John von Newman. He was the discoverer of modern computers, you know, that's what made him very, very famous. von Neumann was asking the question of the details of the measurement problem. How we measure a quantum object which is a wave of potentiality that upon measurement, becomes localized into a particle. How exactly does the measurement take place? What causes it?

Some physicists were ready to give up the idea of causality itself, which ever since the work of the philosopher David Hume, all scientists have taken as sacrosanct. Really, to produce an effect, you have to have a cause. Even early researchers of consciousness, the mystics knew that!

Von Neumann did not believe the idea of effect without cause. He was able to prove a most wonderful theorem, an amazing theorem, a big surprise, a big blow to materialist thinking. No material interactions can ever change a quantum object of potentiality into an object of actuality. In other words, material interactions cannot transform waves into particles or to use the physicists' favorite jargon, cannot collapse waves into particles; just not possible. The causal agency for collapse has to be nonmaterial!

This is the first time in the history of modern physics that nonmaterial objects had to be introduced in physics. And since physics is the fundamental of all sciences, since molecular biology has conceded that biology is chemistry is physics, this idea is part and parcel of how science can be extended to a new paradigm that can explain consciousness, that can explain life, that can explain all things of our experiences left out by the materialist approach.

Behold! A nonmaterial agency has to be introduced in order to make sense of quantum physics, quantum measurement. Quantum Physics without a theory of measurement cannot connect to experimental data. It is not a science if we leave experiments out, it is only theory, half science. Science is theory plus experimental data. Neither experimental data can define science, not theory alone can define science, we have to have both but in order to make sense of the experimental data, we got to understand how the experimental data is produced. And in order to understand that in the case of quantum physics we must postulate that there is a nonmaterial entity somewhere in the measurement process.

Von Neumann gave a tentative answer. It's not even hard to guess once you hear it. Of course, in every gathering of experimental data, there is one common thing, and that's the observer, the experimenter. So, came the observer effect! For every measurement, an observer is involved and it is observer's nonmaterial consciousness that is the agent of collapse.

Say the observer is measuring an electron's wave of possibility, a state of many possible positions. The observer chooses out of all the possible positions that particular position where the electron collapses. The observer's self-consciousness is producing collapse of the wave into a particle.

How revolutionary is this concept? Huh? Before von Neumann, nobody had any idea how to produce a science for consciousness, the subject of pole of conscious awareness, the experiencer, and now we have an avenue because we are introducing observer's self-consciousness into the picture of quantum measurement.

That's the only way we can understand quantum measurement. There is no other way. But Von Neumann's theory still has a paradox. The problem is not solved yet.

Okay, you observe a certain situation of collapse. But suppose a friend of yours also wants to observe. Let's make it concrete. We are looking at an electron, the electron is unmanifest. It's a wave of potentiality. And you have a made an experimental arrangement by surrounding the electron with Geiger counters a whole three-dimensional array of Geiger counters. Von Neumann says you are choosing the particular Geiger counter where the electron will manifest in a given experiment. Unbeknownst to you, there is also another observer who simultaneously measuring the electron, and she is choosing a different Geiger counter, What then? whose Geiger counter will get the nod, will tick? Is there any criteria to decide that? This is the paradox: who gets to choose? Obviously both cannot get to choose, the object is not going to show up in two places at the same time. That's the whole point. It shows up only at one place and who gets to choose? Von Neumann cannot answer. If you think, you are the only conscious being there is, that is the abominable philosophy called solipsism, not science. So, the theory was discarded.

For many decades, many other theories were tried and nothing really worked. Every theory seems to have a problem. In the meantime, a maverick physicist, Fred Alan Wolf, did a wonderful thing with von Neumann's idea. We like mentioning his name because he created this wonderful slogan. "We choose our own reality." He took Von Neumann seriously; he had a feeling that the paradox will be solved one of these days, and the basic idea is solid. It's consciousness that collapses quantum objects of possibility into particles of actuality. We create our own reality by choosing. We have to understand the nature of consciousness to deal with the paradox.

Indeed, the psychology of the question of choice, do we have free will? makes the slogan very interesting. Here, people were feeling doubtful that we even have choice. Materialists were saying, we don't, we are material brain, a cognizing computer. People, we call them behaviorists, were saying that a human being is just conditioned behavior, no being there at all. We don't have any freedom of choice, we don't have any free will, we're mechanical. And here are John von Newman and Fred Alan Wolf who are saying, we choose our own reality because our self-consciousness collapses the quantum possibility wave into actuality. We choose the actuality, and we can choose actuality in favor of what we want. Why not? We can make our intentions come true. Sounds wonderful. Now we are really giving consciousness some tooth. Not only are we conscious, our power, our personal power derives from that self-consciousness, personal power that gives us success or failure, disease or health, happiness or misery.

Is that really true? Can it be true? If that is true, then why can't we just do it? Some people said, well! you are not patient enough; you have to sit and wait for it. Movies were made. You may have seen such a movie called *The Secret*. They say, the secret behind manifesting what you want is waiting; intend it and then wait. Some people even developed meditative techniques to make your intention come true. Since you are waiting anyway, why not meditate, meditate on what you are wishing for. And maybe it will come true.

In seventy's America, many people were working on such ideas to manifest objects of their desire, cars, vegetable gardens in desert areas, what have you. Many claimed success. But many people also did not get success.

Amit's story: I was one of them. I attended a workshop; I was curious about this phrase that Fred had constructed. I didn't know him at that time. So, I went to attend a workshop in which they told us participants to find parking space for our car in the busy downtown area. They said to manifest a car is also possible but that takes too much patience and too much time. "This is a simple enough manifestation and you should be able to do it. They told us to drive to downtown Portland where the workshop was taking place and find a parking place by manifesting it by wishing it and meditating on it. I went an hour early for four days and tried to manifest a parking place for my car. Never succeeded. Always ended up paying parking garages. Whereas I could have saved all that money by taking a bus! that kind of pissed me off, but okay, anything for the sake of science, right? I did end up a loser, and I thought oh, this is this is not an idea that works.

But then the idea would not leave me either; something in it was very attractive. Of course, it would be nice, it would be nice if we had some power to disturb the universe, I did believe that we did have some power, I did believe that we have for example, the power to create, creativity. So, now I became interested in finding a quantum physics in which that paradox against Von Neumann's idea can be resolved.

Eventually, after ten years of research, that solution came to me. I was talking to a mystic, the mystic said something and all of a sudden, I thought, why not? Why not make an about turn from the rational way and think in the mystical way? The idea immediately came to me: consciousness is the ground of all being. You know, my belief was very strong, one cannot do science without objectivity, without the exclusivity of material objects and all that dogma. As soon as I threw the dogma away, as soon as the revelation came, the insight came, no you can do science here, just be intelligent.

Here is the summary of Amit's insight. *Consciousness is the ground of all being and matter consists of quantum possibilities of consciousness to choose from.* That one idea leads to complete understanding of consciousness in all its facets. This is why we are writing this book. So that you get a feel for the treasure that each of us human beings are, that we are conscious beings. And we have all this hidden power to choose provided, we understand and access this place where the choice takes place, the place where consciousness is One, the place where it is the ground of all being, this place we call the domain of potentiality.

Does the paradox, who gets to choose disappear? If it does; how? Because the two people are not choosing with their individual self-consciousness; individual level of consciousness does not choose. Choice happens at a level where everybody is one, from Oneness consciousness, the domain of potentiality. The commonality of intention has correlated both people to become one. And that oneness is choosing, oneness is choosing the place where collapse happens, it's not the choice of either this observer or that observer, but the choice of that one ultimate observer behind every measurement.

We are still leaving out something! Consciousness in Oneness creates reality, but it creates reality through us, through the observer effect. In this way, the observer does play a role. When you learn the details, you will see that you the observer is no less than a co-creator, your exploration will involve creativity and you play important role in the creative process.

This is the story of the quantum. and this is also the story of consciousness. How far does the story go depends on you! Obviously, this oneness you don't access except for those moments of expansion of consciousness that we started our discussion with, that is your hint. If you take those expansion experiences seriously and go exploring all the way and expand and expand and expand, you will be surprised how much of your potentiality you can bring into actuality. The more you are able to live expanded, the happier you become. You will have discovered the endless well that Hindus call Ananda, unbounded joy. As happiness enters your life, you will want to develop intelligence to choose happiness.

In this way, life truly becomes the gift that it is. We'll take it further into the realms of higher consciousness, this awakened consciousness how to live with it, how to incorporate it in our daily life in the rest of the book.

Back to Consciousness

What is consciousness? Okay. It has been confusing. It will take some getting used to and contemplation on your part, some comparison with your own experience, before understanding dawns.

You are not alone in your confusion. There always have been at least two worldviews about all this, it's not just today; this polarization, confusion, has been there, all through the ages. And in fact, the saving grace is that there is a third one. For millennia, there has always been a third one. Let us tell you a little bit of the history.

See, our experiences are of two kinds; nobody can deny that the outer material experience is very different than the inner experiences of the psyche. Mind—thoughts—is the main inner experience, but there are also, the feelings, the emotions. And they are all very different. The outer is stable; the inner changes rapidly.

So, the idea that there are two qualitatively very different substances in the world and they got to belong to two different worlds. But this is dualism, this idea that that there are two dual worlds. One is the world of substance that makes up our psyche, also called subtle substance. There may be more than one subtle substance. And the other one consists of material substance. And the problem here has always been the same, How do the dual worlds interact when there is nothing in common between them. Ancient people could not see the way, modern science could not see the way until recently with quantum physics, with the discovery of nonlocality. The interaction between the two worlds can be mediated by consciousness, using nonlocality, signalless communication.

But of course, years ago, we did have clear enunciation of a monistic philosophy anyway by the sheer power of intuition. This monism was based on consciousness: consciousness is the ground of all being. Consciousness is related to our experience of awareness. We see ourselves as the subject to objects of the world. I and the world, self and the world. But seeing it this way without confusion requires a lot of meditation. To experience consciousness that way, which is oneness, requires creative insight. This is what these guys were postulating, that the I and the world is split as a secondary effect; primarily, reality is oneness.

Except for a few people, people just do not have the intention and perseverance to know the truth, the creativity to get through from separateness to the state of oneness. So, gradually, that particular way of looking at the world, which is today called monistic idealism, became esoteric, sacred, accessible only to a few. Rest of us live in the mundane; we can only worship the sacred from a distance or approach the sacred though middlemen—priests and such—organizers of religion.

The middlemen of religions have no more curiosity than you for exploring the I-experience separate from the objects at a deeper level. But then, where would they put this consciousness as the ground of all being? The way they rationalize it. In this way, they made the Oneness into a personified Being called God and the original mind-matter dualism became a God-world dualism.

In monistic idealism, there is also the idea was that the I-experience comes from Oneness; in the new parlance of God, that would mean that we are somehow part of God, children of God. The more time went, the more the religious leaders had difficulty of recognizing that they are part of God because of their negativity, because of their limitations of me-centeredness; they could not see themselves or any humans thal elevated, enchanted way. And so, the concept of sin entered, we are sinners. Maybe even original sinners; always have been sinners, nothing but. Religions picked up a lot of baggage like that!

Then came modern science. And practitioners of science had a very good idea. Verify all hypothesis with experimental data. The technology was ready to put some of the religious ideas to experimental test. Modern science in this way eliminated simplistic God-cosmologies like the Aristotelian one.

Unfortunately, experimental data is done with experimental apparatuses and these apparatuses are made of matter; you cannot make an apparatus with mental stuff, stuff with feelings and emotions to measure stuff of the psyche except for beings like us. You can also make a measurement yourself, via experience. Here unfortunately, modern science made a grave mistake in adopting a principle called *strong objectivity*. Experience is not allowed as a way of measurements in science, they are subjective.

And that became a very insurmountable difficulty ever since. This limitation that everything has to be experimentally measurable by material instruments only, makes it difficult for making the human psyche a part of science or especially, something like consciousness, or subjecthood.

The idea of dualism could not be justified anymore under critical examination of science. How would two dual bodies interact without exchanging energy? Material bodies interact only with the exchange of energy; energy mediates. But energy does not mediate the interaction of matter and God; *the energy of the material world is always a constant*, no energy goes from material world to God and vice versa. Hence, a dualistic God is disapproved by science; matter is the only reality begins to make sense.

So, gradually, scientists started favouring the idea that matter is all there is. As scientific theory started improving, scientists started applying the materialist ideas, even to biology. And then even to psychology in the twentieth century.

Freud's theory of the unconscious revived a little bit of the idea of consciousness, but there was behaviorism very soon after that, and that became a competition to Freud theory. And the consciousness aspect of Freud theory was soon overlooked.

World War II was an experience of huge constriction of consciousness for all human beings; Americans were not exception. The veterans of the war wanted to settle down with old conservative ideas, but their children were ready to move on. Gale winds of change were everywhere: the hippies, birth control pills and sexual liberation, the Women's lib movement, Martin Luther King, TM, the Beatles, Zen, LSD. Together these ideas and movements threatened to expand consciousness, like never before. America was going to lead the re-enchantment of the world, as the rest of humanity were ready to follow,

Did scientists of the establishment become kind of desperate, enough to declare that everything is matter, premature as it clearly was in hindsight? Did they know that they're not going to be able to explain human consciousness or spirituality, or archetypal values with the hypothesis of material monism ever because quantum physics was already ruling that out, even with the limited understanding achieved by then? Only history will judge why they did it.

Of course, the political parties today are very confused. people today are very confused because of the worldview polarization that the scientists' decision led to. Where do the human values of democracy come from? An ambiguity is created opening the door wide for hypocrisy. Today, one party in the USA thinks that the values are only a pretence. They impose values on people as the politically correct, acceptable behavior. You should do this; otherwise, how can there be civilization? But of course, simultaneously, they believe in a materialist science which does not uphold the values. The other party are just as opportunistic and as much of hypocrite. They pretend to support religion, but don't follow the religious values anyway. And on top of it, a substantial portion of both parties' voters are actually quite religious; they follow religious values seriously in spite of all the ambiguities and the chaos created by the leaders. That's how the culture of the country somehow survives. This is the story of politics virtually everywhere.

And there is also that 15%, I keep mentioning. Strangely, the spiritual tradition, the wisdom tradition, the esoteric traditions of religion, they never completely went away. The Vedanta philosophy remains strong in India. Not too many people pursue it, but a few people do take it seriously and produce enlightened masters on a regular basis. Buddhism is very strong all around the world in following their wisdom tradition. Yes, popular Buddhism is what adds to the numbers, and the followers there, too, are a confused people. But there are many philosophers among Buddhists, they go into the deep questions, they meditate, they believe in darshan. In Japan, Zen was very strong, until the World War II. in Eastern Europe, if you go to a country like Romania, you will find lots of spiritual people. Same thing is true in Italy, in Southern Europe, in Brazil, in Kuwait, lots of people who together make up this 15%.

In this way, we really should always think of the world having three worldviews: one a spiritual minority, very much of a minority. The other two are quite strong and very vocal. And those are the two we hear most! The political parties and the media align with one or the other, but nobody believes values. At least, there is some recognition that values are necessary for civilization, we must at least pretend but many people don't like pretension. So, we have a mess. Either we choose inauthenticity or we choose hypocrisy. The latter is pulled off by faking authenticity itself; it requires good acting. In this way actors are succeeding in politics.

The few people who try to understand reality, they too are stymied by the question: how do we incorporate spirituality and values in science? How do we show scientifically that indeed, there is consciousness? It is in this perspective, we have to understand the impact of the quantum worldview which has been developing ever since the nineteen eighties.

When I (Amit) myself and Berkeley physicist Henry Stapp and a little later, Rutgers professor Casey Blood came up with this idealist consciousness-based interpretation of quantum physics, where indeed, nonmaterial consciousness is recognized as the ground of being and experience is valued, experience happens when consciousness chooses out of its possibilities the actual event. And consciousness can choose all our experiences that way, nothing needs to be demeaned or discarded. And therefore, other nonmaterial entities also now can be included in a new paradigm of science called quantum science. The results have been so powerful, the agreement with experimental data has been so powerful and convincing that even the great religious spiritual leader Dalai Lama today thinks and he has said so publicly, *that no model of reality is possible without taking account of quantum physics.*

Riding on quantum science's back, we have integrated science and spirituality. We have even verified the main concepts. Is our brain quantum? Our theory says, yes. Does it have quantum non-locality to embody a nonlocal self-experience? Yes, it does.

Amit's story: I'll tell you the experimental data that shows that yes, the brain can be nonlocal, it does have observable quantum behavior. The experiment was the work of a neuroscientist named Jacobo Grinburg. When I first discovered this idea of consciousness is the ground of being, I wrote a scientific paper in a physics journal. This guy Jacobo, Mexican guy, Jacob in English pronunciation, he called me and said, I have some very interesting data, which you might like to see. So, I became curious, he invited me to come to University of Mexico in Mexico City. I went and I actually observed the experiment and interpreted the data, when it was analyzed.

Here is how the data was collected. Remember, nonlocality works only with correlated beings, objects or human subjects. And what causes correlation between material objects is that they just need to interact, they come close and interact. How do you corelate two human beings? this was the first challenge and Jacobo figured out the answer. He had some experience with meditation. Meditation is known to have effects on people's brain. You know, this, quantum comes with an idea called coherence. You have seen brainwaves, right? They are the average vibration of the different parts of the brain represented in terms of these waves of various frequencies that we call alpha, beta, etc. Jacob found that these brainwaves, if people meditate, become somewhat coherent between the back and the front and the sides too, of the brain. Different parts of the brain seem to be a little more coherent in moving with one another. Now what does the word coherent mean? Coherent is very important word for waves. When two waves dance in step, like dancers in a Chorus Line, we say they dance in phase, that's the more technical way of putting it. But really, then the dance is coherent. Two objects, which are vibrating in phase, in perfect step with each other, they are coherent. So, what this means is this and Jacobo was the first to notice it, but later on this was also verified by fMRI experiment, functional magnetic resonance imaging: the more we meditate, the more quantum the brain becomes, between the different parts. They become somewhat synchronous in movement, coherent with each other, quantum.

The idea came to Jacobo to let the two subjects meditate with the intention that they will have direct signalless communication, to achieve correlation between them. The protocol of the experiment went like this. Two subjects meditated together in a Faraday chamber with the intention of direct communication during the whole duration of the experiment. But the initial twenty minutes or so, they meditate in proximity; after the initial twenty minutes one of them was moved and put in a separate Faraday chamber. Faraday chambers are rooms through whose walls no electromagnetic field or wave can penetrate. So, the subjects are isolated as far as any practical signal is concerned. The subjects continue to meditate. Both subjects have individual electroencephalogram machines connected to their brain; these machines can measure the electrical activity in each brain. Okay?

So, they're meditating, the two subjects; one subject now is shown a series of light flashes Why not one flash; because you have to maintain this state of affairs in the brain for a while. Otherwise, nothing concrete can be concluded; we have to take the average of the brainwaves, work on the average. So, one subject sees a series of light flashes and that subject of course has all this electrical activity in his occipital cortex and eventually all the involved areas of the brain and that is recorded in the ECG machine showing electrical activity that is summarized as a signal called evoked potential. You have to eliminate the noise from the raw data to get the evoked potential; the other subject is not seeing any light flash. This is the crucial information. The other subject is just meditating with the idea of oneness in mind with the intention that there will be oneness between the two, they will be able to communicate without any signal just because they are meditating with an intention.

And lo and behold! In Jacobo's experiment, the EEG machine of the other subject started responding and showed electrical activity; when the noise was eliminated, the signal was truly a big surprise. The signal which was called transferred potential, had virtually the same phase and the same strength as the evoked potential. *But the other subject had not seen any light flashes.* What explains the transferred potential? She was sitting in an electromagnetically isolated chamber, how could electrical activity go from one brain to the other except for nonlocal transfer through the medium of consciousness itself.

Okay, control subjects, that is very important aspect of this kind of experiments, control subjects, subjects who did not meditate, what happens for them? Well, they did not show any transfer potential, never. Meditation is crucial; intention was crucial. Control subjects who did not meditate at all or could not maintain the meditative intention never showed a transferred potential.

Now initially, the reaction of people was: we don't believe you. I remember I came back to the physics department where I worked and shared the paper we wrote with the data with my dear friend, very intelligent physics colleague and he said, Amit, you must not put your name to this paper. He said this will completely ruin my reputation. Of course, I didn't listen to him and he was very sad for me. Guess what? five years later this experiment was replicated by a British group. And then again, another group and then a group in America itself. And now, there are at least two dozen experiments in different laboratories of the world by different groups of scientists that have verified the same transferred potential. Yes, electromagnetic information can be sent from one brain to the other without any physical connection, via quantum nonlocality through the intermediary of consciousness, proving beyond any doubt that brain is quantum and consciousness indeed, can mediate such interaction which previous models would reject as dualism.

What kind of monism is this? It is monism based on consciousness—monistic idealism, consciousness is the ground of all being. Finally, we have a better description, a better understanding.

But the Oneness really belongs to the domain of potentiality. In the domain of actuality, when we have experience, then this oneness collapses into a subject and an object. The subject-experience arises from the identification of consciousness with the brain, and the object is experientially separate from it. That's the way the perception goes. Awareness has two poles—subject and object, I and the object.

Our confusion comes because normally we do not experience this nonlocal quantum self; what we experience is what the brain deals us after a series of reflection in the mirror of the memories it makes of every such experience. The ego experience is the result of dependent co-arising of me with the I practically gone implicit and the object in its conditioned familiarity.

What does meditation do is to take us towards the quantum I towards the Oneness; done regularly, it establishes more coherence in the brain and takes us into a domain that we call preconscious, in between the ego state where we have almost totally me-centeredness and the I-state where we are fully in the quantum self (*no-me*). So, meditation is very, very important to get a grip on the nature of consciousness. That's one thing.

Second thing is that all human experiences can be explained and incorporated in this new science. Now we can talk about feeling, we can talk about meaningful thinking, we can talk about even the experience of intuition of the archetypes, values, spiritual values, that can change us, that can transform. In the new science, we can have both creativity and conditioning. Because the unconscious which is the domain of potentiality, not only has the conditioned personal part that Freud discovered and the collectively conditioned part, memory of the whole humanity that the psychologist Carl Jung discovered, that is called collective unconscious but also the previously un-collapsed possibilities of consciousness, the quantum unconscious, we are able to collapse that which is new and creative.

And now you will understand that phrase that Fred Alan wolf, the physicist who introduced consciousness to the people, via the slogan we create our own reality, and appreciate it fully. How do we create our own reality? we create it by delving into the quantum unconscious creatively. We get Oneness on our side. In old parlance, we get God on our side. Its oneness that chooses; if we can get to oneness, if we can access oneness, we can also choose anything we like. That's how we create our own reality.

Can we have good health? Yes, we can. Can we have happiness? Yes, we can. We can choose provided we meditate, engage intuitions and creativity, and discover our identity with this oneness. The door to the Oneness is the enabler: whenever we ask, can we do this? Can we do that? Yes, we can. Anything permitted by the science that the One created the universe with. Only if we can access that Oneness.

Otherwise, how do things happen? They happen because conditioned actions interact with conditioned actions; the result is other actions but they are all determined actions determined by the past. Of course, there are lots of human beings with lots of complication. So, we will never be able to calculate the outcome of such interaction of conditioned actions with conditioned actions. So, things may seem a little indeterminate. But, even so, some of the actions can be predicted by looking at people's common habits and stuff. So, it gives us some stability, some predictability, agreed. That's what behaviorist science is about, this is what scientists like. But they have lost much because of their preference of "predict and control." They have lost, and with them you, have lost the idea of enchantment—that feeling of connection that we have with the world, automatic when we are children. Is it worth losing that privilege to a belief system?

Is it worth losing creativity? Is it worth losing meaning? is it worth losing the values, we are finding out more and more that it is not worth it. Without values, it is chaos, and we cannot live with chaos. If nobody has any values or even pretend values, we cannot have society, we cannot have civilization.

An American election happened on November 3, 2020. Is there any lesson for anyone? It is that people cannot live without values, it is too much chaos. Pretend values will not do enough; people have to discover values, transform, and use the values in their life. The archetypes—goodness, love, love, truth, justice etc. have to come back in our lives, in our society.

What can bring you back love? Is love scientific? Let's look at love from the perspective of quantum physics. Two people interact, local interaction, nothing much happens. But if nonlocality enters, the two become one. Ordinarily, I see you as an object. But when I'm in nonlocal connection with you, then I see you also as a subject, when I collapse from the state of Oneness. Now I can truly love you as your equal, I recognize that you are also a subject not an object for my pleasure, not an object for my whims to obey me, and vice versa. Instead, we recognize that we both have equal footing, we are both subjects. Out of our volition, we are agreeing that we will correlate and stay correlated as much as we can. Of course, events of the world will separate us occasionally. But we know that it is love that will sustain our correlation, sustain our relationship.

Of course, people can choose not to correlate with other people. So, some people will do that and therefore, there will still be division, there will still be a lot of separateness. But if a substantial number of people discover nonlocal love and change their lives, the entire society can change. We can create a quantum society. How much of the quantum can we bring and how soon? Our entire civilization, the future of humanity may depend on the answer to this question. What is your decision? What is your choice? Love or separateness? This is the question for you to decide.

CHAPTER 3

The God Controversy

When I (Amit) was growing up in India in the 1940's and 50's, God-talk was quite common among young people of all kind as was the talk of morality. But things are quite different now. The middle class almost everywhere think of the subject as taboo except those growing up in religious families. This exacerbates the science-religion divide.

From the middle-class perspective, religions have become more militant, more fundamentalist. If you point out that this is a direct result of the assault, a direct threat of annihilation that materialist science has put on religious beliefs, most people would be surprised that you are using the special label of materialist science. Isn't science the same thing as materialist science? they will ask. They don't know that science with a dogma is not real science; it is borderline pseudoscience, quite credible for its mostly useful empirical data, but its theories are dubious whenever they deal with the human being.

And if you counter with the retort, Is religion all there is to spirituality? again most people will be surprised. They are not able to discern between the religion and spirituality. Religions have dogma; the theories of spiritual wisdom traditions are based on data, human experience of not of one person, but many. Just like much of cognitive psychology data collected today.

Is the research about the nature of reality a game in which turnabout is always fair play? When modern science began in the West with Copernicus' discovery of the solar system that replace the earlier earth-centric model of the universe, religion—Christianity to be specific—fought tooth and nail with the emerging science.

Christianity had made major investment in their own earlier model based on Aristotle's science, how could they be expected to do otherwise? Inertia, power game, both are a part of human nature.

However, the reaction was barbaric, burning scientists at stake for example. So, the question can be asked, Are scientists now reciprocating that antagonism when they reject God and moral values with vehemence?

Our point is this: materialist science is not science just as religion is not spirituality; both are dogmas imposed upon two legitimate tracks of human inquiries about the nature of reality; it is the philosophy that divides them. Science is an inquiry into the nature of reality based on theory and experiment; the earlier spirituality is the same inquiry based on theory plus experience. The former emphasizes objectivity and our outer experiences; the discovered truth should lead to technology that enhances the external life—survival issues—of humans? For the later, the criterion is livability—can the discovered truth be lived? The idea is to improve our internal life and relationship with others? Both approaches are necessary for growing civilization obviously.

So, we have to examine the God-controversy from this context. A context of finding common ground rather than division. Then the resolution. But first things first! Many readers are ignorant of the history of how the divide seen today came about. Let's fill in the gap.

A Brief History of God in the Mind of Western Explorers

Modern science really began with Galileo with his demonstration that the satellites of Jupiter revolves around the mother planet in the manner Copernicus proposed. Then in 1665, Newton theoretically proved that his gravity theory applied equally well to the movement of the moon around the earth as it did for the movement of a falling apple on earth. Newton's theory of universal gravitation was verified by lots and lots of experimental data in due course.

This changed history. Why? It is the rationalism of the Church fathers! The Greek philosophers Socrates and Plato independently discovered the monistic idealism version of the nature of reality about 2600 years ago. Reality is Oneness Consciousness breaking up into subject-object duality for experience. Oneness is transcendent; experience is immanent.

But this was too abstract for Plato's brilliant student Aristotle who made it "real" by theorizing that the transcendent reality—heaven—consists of the heavenly objects of outer space, the moon, the sun, the stars. Obviously, heavenly laws are different from earthly laws to reflect the former's perfection. In due course, Christianity bought into Aristotle's theory giving up the idea of nonrational transcendence in favor of something that everybody can experience and admire with awe. The triumph of rationalism over the nonrational.

Newton demolished Aristotle's simplistic picture. Popular Christianity lost one appealing popular concept—God sitting in a throne in outer space. Where? God is omnipresent, so everywhere really. This is why that joke about the importance of Neil Armstrong's walk in the moon and not finding God being the turning point in favor of materialist science!

Joking aside, it took no less than a hundred years to shake the scientists' belief in God. Well in the eighteenth century, emperor Napoleon was curious about a new book on celestial mechanics. So, he summoned the author—Pierre Simon de Laplace. "Monseur Laplace. I hear you have not mentioned God even once in your book!" To this Laplace famously replied, "Your majesty! I have not needed that particular hypothesis."

And then, the genesis picture of creation was compromised to a great deal with the publication of Darwin's book *The Origin of Species*, published in 1857. But still the Cartesian truce held; God lost authority of the material universe; but her authority over the mind, morality and all, was deemed unshakeable.

In the mind of many rational believers the religious worldview underwent quite an effective rationalization, however. God created the universe; and then retired. Seriously, except to occasionally dabble with people's mind. This philosophy called Deism was believed by a lot of scientists including the great Einstein.

After World War II, the cold war and the competition of space travel, changed the practice of science. Overnight, scientists began do get big research grants. Most of it went to weapons research but a substantial amount did come to the universities in the USA. It is that classic story of a Faustian selling of soul that the famous Goethe wrote about.

In the nineteen fifties and sixties, rationalism in America was at a crossroads. The Easterners had come in bunches; the Hippies had declared war on "tight-assed" rationalism, beetles had popularized meditation by embracing TM, LSD had cracked the myth of rationalism wide open. Women's lib was threatening with women's values—emotion—and was challenging white male rational domination; the civil rights movement threatened white agnosticism, blacks were devout Christians. The psychologists were threatening with human potential movement and transpersonal psychology; Traditional Chinese Medicine made an entry in America in the form of acupuncture.

Even the theologians joined in. Many decades ago, Nietzsche said, God is dead. The saying became part of a new God is dead Theological movement which declared the death of Christian rational theology.

The choice was clear. Give up rationalism and accept the super-rational or grab the opportunity to kill popular Christian dualist Theology.

The scientists needed a bold decisive step and chose the latter idea. They claimed that reality is matter, matter is the only reality. That was the genesis of materialist science. Matter is everything.

Richard Feynman expressed the desperation of the scientists to take the step well when he said something to the effect that scientists must wear this straitjacket of materialism in order to do science, take science further.

The dogma, correctly labeled as straitjacket, killed, suffocated not only the God of Christianity but all metaphysics, including the Eastern, without even examining the latter. Along with killing darshan, it killed truth itself. It killed all the archetypes, love, morality, beauty, justice, wholeness, self. Consciousness itself.

These materialist scientists jumped the gun. Materialist science served their greed and the will to dominate society, but in its aftermath, after some decades of it, who can deny that what followed—the worldview polarization and the crisis of democracy—was too great a consequence to pay? Civilization cannot survive without absolute truth. Dogmas, by killing the nonrational, open the door to the irrational, the conspiracy theories that now dominate the Internet, which itself is a crown jewel of modern technology.

It is likely too late to save the earth from the consequences of human-made climate change. It may likewise be too late to bottle up the genie of irrationality. But we must try. With quantum physics, downward causation and the three principles of nonlocality, discontinuity, and tangled hierarchy, let's declare materialism is dead. Don't worry! Just the philosophy. The science discovered with its help that has been verified stands although the interpretation of the data will change in some cases. Likewise, the useful material technology stays. But the absurd metaphysics has to go to liberate science from its shackle. Promissory materialism will have to stop; it is not going to deliver.

God: The Comeback Kid

Yes, before your very eyes, God is being reborn, the concept that is. How should you think of God in view of the fall of materialism? In chapter 1, we commented on how the concept of God is incorporated in monistic idealism that I (Amit) showed quantum physics fully backs up.

But the solution of the quantum paradoxes that I proposed was immediately criticized as "going too far" or "too Hindu" in flavor.

Rationalism dominates the post-materialists, especially the hard scientists among them, no less than it does the materialists. The quantum theory of consciousness was based on physics, biologists complained. Why can't biology take the lead here? Or psychology?

Fortunately, avant-garde psychologists had already developed transpersonal psychology. Under the leadership of a young talented maverick, the philosopher Ken Wilber in the late seventies and eighties, psychology went spiritual. Wilber firmly established the idea that transpersonal psychology must be based on perennial philosophy which is another name for monistic idealism, more or less. Unfortunately, later Wilber himself, being a philosopher found a new pasture and moved on. Fortunately, transpersonal psychology ever since has accepted the spiritual ontology.

In case you are curious, Wilber developed a phenomenological approach to science of consciousness that purportedly integrates the personal, transpersonal, and objective—I, we, it, and its— aspects of our experience. Unfortunately, his four-quadrant theory is just the old mind-body dualism in a different guise.

God is not needed in Wilber's theory. The "we" consciousness that brings in the transpersonal is a local "we;" without nonlocality, it misses the spirit of transpersonal psychology.

Wilber adapted the idea of holon—irreducible wholes—from earlier work; but there was no scientific demonstration of the concept. In Amit's work, the only mechanism that produces irreducible whole is tangled hierarchy, but this is only realized in living matter. Memory is impossible without the participation of matter.

The physicist Fritjof Capra of *Tao of Physics* fame, resorted to systems theory to explain life following an approach by biologists Humberto Maturana and Francisco Varela. The latter authors attempted to produce an irreducible whole with the intriguing concept of autopoiesis—poetry making. But their demonstration of irreducibility was incomplete.

The concept of holon, autopoiesis, etc. are all part of a philosophical approach called holism defined simply as "the whole is greater the parts." In this way an illusion is created that the irreducible wholeness" of holons and autopoietic systems are examples of this emergent wholeness, the "whole" that is greater than the parts. However, in the absence of any viable demonstration of such an emergence in an actual manifest system, it is all pseudo-rationalistic philosophy.

Holism does have room for God; it is called pantheism. God is the "wholeness" of the material world itself. The concept of a transcendent God is given up.

In another line of thinking, the physicist David Bohm found an approximation of quantum mathematics. He demonstrated two important ideas to make a case for God in a new way. 1) There are two levels—Bohm called them orders, implicate and explicate—of reality. This is analogous to heaven—God's abode and earth—human abode. 2) The implicate via a nonmaterial and nonlocal interaction which Bohm called quantum potential, influences the movement of an object in the explicate that we see and measure. In this way, Bohm mathematically, albeit in an approximation to quantum equation proper, demonstrated that there is downward causation

Notice that Bohm's theory is deterministic; what seems like indeterminacy comes from the effect of quantum potential on the explicate movement. Exact trajectories can always be calculated after the effect is known. Bohm's God is thus quite compatible with the philosophy of Deism—after creation, God becomes benign, leaving the world to evolve as per Bohm's equations.

One idea of spiritual wisdom traditions that has always caught popular imagination—and it should—is that humans are in some limited way God. Jesus said, Ye are all God's children. That's one way of thinking. A more popular way is that we are little "rays" of God. This idea has reverberated in post-materialist thinking in the form of the holographic theory of the universe.

Hologram is a laser-enabled photograph of an object with the special characteristic that even a little part of it retains the basic info of the original although with bad resolution of the details.

You see the analogy. If the universe is holographic, then even little parts of it will in some way have the knowledge of the whole in the same way as we are made in God's image retaining bits and pieces of Godliness.

There are other lesser-known theories of this ilk that are cycled and recycled in an ongoing way in the new age universe of post-materialist rational thinking universe. Some are overtly materialistic like the complexity theories in which emergent phenomena such as life and consciousness are epiphenomena of complexity of the material assembly that we are dealing with. Or chaos theory. It is a fact that Newtonian determinism is only one of principle. In practice, very few systems can be subjected to accurate deterministic calculations. It is also a fact that a little change in initial conditions often produces a big change in the outcome. In this way, the actual movement of complex systems, although deterministic, may end up in chaos, determined chaos. The curious thing is that sometimes unexpected order can emerge in the chaos—order within chaos. These could easily be theorized as the "surprising new" that we have previously identified as the products of creative quantum leaps. Of course, such a theory leaves *you* no role to play; it is all in the chaos dynamics of your brain. Your quantum leap of thought, your subsequent experiences of the wonderful flow of manifesting your creative insight are just ornamental epiphenomena while the brain is doing the real work!

In this way, the idea that human creativity is a demonstration of human connection to a deeper cosmic consciousness has been challenged. Here again, if you accept Bohm's approximate version of quantum equations as gospel truth, then, a Deistic benign God or some sort of benign Oneness consciousness (after all Bohm's implicate order is nonlocal, the quantum potential is nonlocal) can stay.

The common thing about all these models is that they are rationalizations; they are claiming to take us beyond the limits of materialist science and doing that are providing plausible excuses to hold on to the basic idea of material monism or realism or both. Their limits are also the limits of continuous logical step-by-step rational thinking. The sad thing about this is that you can go on and on with such imaginative adventures of rationalism but never find any answers that requires going beyond rationality. Unfortunately, they keep post-materialist scientists busily in search of that one Nobel-winning theory someday.

Philosopher Noam Chomsky has a name for these mainly Western post-materialist intellectuals; he calls them intellectuals with values. This is still something. But the data demands more; it deserves more. Why are these intellectuals so reluctant to look at the data on nonlocality, creativity, vital energy, survival after death and reincarnation, meditation, unconscious, fossil gaps, biological arrow of time, anthropic principle, etc.?

It takes an open mind and about twenty minute of awareness meditation to fall into a state of expanded consciousness or relaxation with proper guidance. Here expanded consciousness means inclusivity; you are able to feel generosity or caring for others. Why is that so hard with your 150 IQ intellect?

In the meantime, quantum physics is giving us answers to all the relevant questions of consciousness and God. Read Amit's books, *God is not Dead* and *The Everything Answer Book* and convince yourself.

Does God Create? The Reconciliation of God and Biology

Biology is the science of life and its evolution and development. The four basic questions are:

1. What is life? How did it come about?

2. Did God create the universe with purpose?

3. Does life evolve and how does life evolve?

4. Is it all evolution or is there also development of life for which the process is different?

What is Life?

Before the advent of science, in the religious worldview, quite generally God is given the credit for creating all life. Confusions abound. God is supposed to be everywhere; and life is one proof of God's potency. Then why is a piece of rock not alive, not conscious? Neither religions nor the wisdom traditions could answer satisfactorily. Saying that God is in a piece of rock as much as he is in a human being does not stop us from asking, prove it?

Modern biology has never explained the difference between nonlife and life either except taking it as an empirical fact and the pseudo-explanation that there is no difference. Biology is chemistry is physics. Matter somehow evolves macromolecules of RNA, DNA, and proteins and voila! Life. But calculations based on probability mathematics conclusively shows it is impossible to put together programmed molecules such as DNA and protein without a programmer. Additionally, the programmer needs organizing fields to do so. Finally, self-identity requires the full cell—DNA, RNA, protein, cytoplasm, cell wall of confinement—for tangled hierarchy using proteins in the cell wall for perception and water in the cytoplasm for memory.

Why isn't a rock alive? Because it lacks the programmed molecules and the tangled hierarchy. Why isn't a virus alive though it has the programmed molecules. Because it still lacks the cell's tangled hierarchy-making apparatuses.

Does God create life? You bet. A programmer is needed; call it consciousness or God—your choice. But a creator is needed, unconscious processing and downward causation are needed.

Is the Universe Purposive, Serving God's Purpose?

Cosmologists agree: There was a big bang as predicted by Einstein's theory of general relativity. Many theologians, even some scientists, have been encouraged by this to say that the singularity of the event proves God, proves God as creator of the universe.

In quantum science, quantumizing cosmology—looking at the creation of the universe first as a bunch of possible "baby" universes, and then invoking downward causation—eliminates the singularity and establishes the proper role of God and downward causation in the creation event. The choice can only happen when God—unity consciousness—can manifest, can identify with a tangled hierarchy in a cellular assembly. That is, with the appearance of the first living cell. In this way, the universe is created retroactively, going backward in time. Such a way of applying choice is called delayed choice in quantum physics.

How to empirically prove that God's downward causation created the universe? Downward causation is purposive, as we have noted before. Is there evidence of purpose in the evolution of the universe over time? There is. Data after data suggest that the universe is created in such a way that life will eventually come about in some parts of it. This is called the anthropic principle.

Reconciling Darwinism with Christianity

In Judeo-Christian religions, in the Book of Genesis (which means creation) it is all laid out in definitive terms: God created all life in six days a mere six thousand years ago. In Hinduism, the original literature of consciousness, life and its associated energy are acknowledged, in this way, that life is different from nonlife is acknowledged as well. (This is true of the Jewish Kabbala as well.) But no reference to evolution is found until the Puranas which appeared much later. In the Puranas, God is theorized to appear in different times as avataras to rescue the movement of consciousness to proceed in the right direction: the first avatara is fish; the second one is amphibian tortoise; the third one is a mammal in the form of a boar, and so forth, suggesting evolution.

Modern biology's triumphs are manifold however beginning with Darwin's theory of evolution. Modern interpreters of Darwin's theory (reformulated as Neo-Darwinism) are materialists. It holds that life evolves from a single cell via a two-fold mechanism:

1. Chance variation: changes are produced in the hereditary material—the genes which are parts of DNA—via mutation, errors during sexual recombination of the male sperm and female ovum, etc.

2. Natural selection: out of these variations, nature selects those variations that have survival value against ongoing environmental challenges. When enough changes accumulate, the new conglomerate of cells cannot mate with the old species anymore; in this way, speciation, a new species has emerged.

Science has two prongs. Theory must be backed up by empirical data. For evolution, this is provided by the fossil data. At first site, the data shows evolution from a single beginning as a tree of life. Closer examination reveals problems:

1. The fossil data is not continuous, that is, the tree of life is not continuous. New branches start abruptly without any discernible continuity with the old;

2. Old species often survives all the climate changes since the beginning of life in one long continuing stasis;

3. Often, maximal number of new creations of life is accompanied by climate catastrophes that destroys a lot of the living species.

Darwinism or Neo-Darwinism have no explanation for these details except non-verifiable pseudo-scientific ones. For example, a portion of a species community can become separated by the appearance of a sedimentary mountain range and is forced to mate in small communities which is known to produce wild mutations. Some of these wild mutations could have had survival value. This theory is called geographical isolation theory. An unveriable theory, no more than a plausible storyline like the Hindu Puranas avatara theory.

On the other hand, Christian rationalizers can point out the following in favor of God's genesis of life:

1. God creates life; different species at different times. The idea of "six days" of creation metaphorically captures this.

2. It is well known that the story of Noah's arc followed a great flood. Great flood is a metaphor for a geological climate catastrophe. Biologists have no explanation for survival of the luckiest. Noah's arc gives a metaphorical explanation.

Ok, ok. We are cheating a little. The actual creationists are fundamentalists, believers of literal interpretation of the Bible. But truly, put in this metaphorical form, is the creation theory of Noah's arc—God's Grace protected the chosen few of the many species from catastrophe any less valid than the geographic isolation theory of the Neo-Darwinists? Neither can be verified. Both smack of pseudoscience.

The serious point is this: Neo-Darwinism cannot explain the fossil data. It is a triumph of consciousness-based theory of evolution (read Amit's books *Creative Evolution* and *Quantum Biology and the Ascent of Humanity*) that the fossil data is fully explained. The explanation asserts that quantum creativity is essential to explain the details of Darwin's theory and when we do that, the fossil gaps are properly explained. Need we remind you that the creative quantum leaps are the result of downward causation, God's interference, so to speak?

Human creativity, albeit the result of God's Grace in the form of downward causation from Oneness consciousness, very importantly involves also human effort. Where is the vehicle via which God's creativity expresses itself in biological evolution? Individual animals do not display any such ability! The answer is group consciousness. In biological creativity, nonlocality adds up to the efficacy needed for immanent effort to bring down God's downward causation.

There is evidence for this in the phenomenon of directed mutation. When threatened with starvation and extinction because food they could digest was not available by the design of the experiment, bacteria, acting in oneness changed their mutation rate to mutate to another species so that they could eat and digest the available food (as per the experimental design). God's Grace, yes but also species' creative effort.

Biological Development: Purpose in Evolution

The purpose of evolution is easy to see in the fossil data. The data has one-way-ness, you can call it the biological arrow of time; the fossils evolve from simple to complex. The complexity is needed to respond to higher needs (than mere survival) of the evolving consciousness—the purpose of evolution.

If we look at it the realist way, purpose sounds like a cause from the future. Indeed, Aristotle called it final cause. Philosophy of purpose is called teleology. It is anathema to material realists.

In truth though, if we are objective enough, it is pretty obvious that as life evolves, the species are able to perform newer functions that serve purpose, albeit the purpose maybe better survival. Take the case of locomotion, the major distinction between plant life and animal life. Locomotion is a purposive function. Animals cannot feed themselves directly with solar energy (which plants do through photosynthesis), therefore, when they exhaust the food in one pasture, they got to move to another one.

Biological function is performed by the proteins programed to do so. Proteins are very complex in structure; they cannot be made by random assembly of molecules. So, the science of consciousness provides a programmed molecule, the DNA, portions of which—the genes—has the code to make proteins from amino acids. For making life, to evolve life, we need programs; a programmer alone will not do, also an organizing field to assist the programmer.

Compare the situation with how you use your computer hardware by making software, the purposive functions that you want your computer to do for you. You use you mind to help your consciousness, right.

So, consciousness uses nonmaterial organizing principles called liturgical /morphogenetic fields to make the vital software to go with the biological hardware —the organs of our body. The biologist Rupert Sheldrake first theorized about the necessity for introducing these organizing fields in biology.

And here is another problem. Darwin's theory tells us about the evolution of the hardware from a single cell to composites. The genes are hereditary; genes are how the changes in hardware propagate. But how do the changes in software propagate?

It is not widely known, but Darwin was not a materialist; throughout his later life, he looked to include purpose in his theory but in vain. He did not have the right context to posit his question. In truth, even the genes were not known yet; Darwin had to use his intuition to suggest that there must be a hereditary molecule.

There was an existing theory of evolution by another biologist who preceded Darwin by the name of Lamarck. He intuited that it is the software that is hereditary!

Darwin's prediction of genes was verified only decades after. No such luck with Lamarck's intuition. Only recently, have biologists verified that there are epigenetic programs that guide the genes to be activated in cells belonging to a specific organ performing a specific function.

So Lamarckian evolution dictates that these software programs are hereditary. Here again, if they were material programs, there would be no way. However, because the software is nonmaterial, it can propagate through the mechanism of nonlocal memory.

The making of software is called development—development of the program. In this way, it is now clear that Darwin got one half of the evolutionary mechanism, Lamarck the other half. Evolution is evolution and development together; let's call it Evo-Devo following one author who anticipated something like this.

This picture is possible only if the movement of consciousness propagates evolution not the phony idea that molecules have survival instinct. Conscious downward causation—choice—can be purposive, obviously.

God's Plan for the Future of Humanity?

Teilhard said evolution will lead to heaven on earth; he meant development of course. Is this God's plan for us?

Let's keep away from the word plan; there is no predetermined plan in quantum science. However, like all science, it allows us to make predictions. Can we predict the future of humanity and does it agree with Teilhard?

Like Teilhard, Sri Aurobindo also had some predictions. He said we will develop into Overmind—connected consciousness—and even beyond Supermind. With enormous new scope and power. Does quantum science agree with that?

Finally, what is the lesson here for you and me? In order to get our intentions fulfilled we better learn to align our intentions with the purposive movement of consciousness. Big lesson.

Let's use the God language. How does the objective scientific language translate into the subjective God language? Aligning with the movement of consciousness is to get God on your side.

Religions for centuries and millennia have declared, We are the chosen people. So have dictators and kings. Have they been? Now we know. Very seldom. The chosen are the people who transform to do God's work, not ego's work. Got it?

Consciousness or God? Wisdom Traditions and Religions: The Medium Defines the Message and It Ain't All Bad

We have been emphasizing one aspect of all this: religions simplify, fall prey to dogma, become exclusive, and divide. Whereas wisdom traditions are of one voice. This is looking at religion from the negative side.

There is another way to look at things, look for commonality. Are their common elements among religions. There are indeed. All religions propound three aspects of reality:

1. There is a higher power, a nonmaterial source of causation—call it downward causation to fit the metaphor of "high." All religions agree to name the higher power as God or something equivalent.

2. We not only have the material body, something akin to what today we call hardware; we also have subtle bodies; what today we call software. The hardware is made of matter; the software is made of subtle objects. The vital movements we experience as feelings; the mental movements as thoughts. Feelings bring us passion; thought give us meaning

3. Beside thoughts and meanings, we have the experience of intuition for which the objects we call archetypes. Archetypes of consciousness. Religions think of them as God's virtues. Intuitions come to us via extrasensory means. Intuitions and archetypes bring us purpose.

Religions want you to make a personal relationship with reality. Why do you want to cultivate God's virtues? Because you love God you want to be like Him/Her. You know the adage: God makes humans in His/Her own image. Not true obviously if we look at our current human condition. The real lesson is: God wants to make you in His/Her own image. You are an unfinished product. You have to develop further. You have to explore and embody your infinite potentiality.

Spiritual traditions long intuited that we have two important selves: one in the brain; this is the dominant one. The other one is in the body, around the heart. This one has to be awakened; fortunately, it is awakened in part in women and also some men. The heart cognizes with love; the brain with thought.

People who are dominated by thought, they want to investigate reality objectively. The people of the heart via relationship. A direct relationship with God. The former gives us the wisdom traditions, the latter the religions.

How do religions take on dogma? It is the leader-follower thing. Today we see it in the phenomenon of Donald Trump and his 75 million followers. What is the most conspicuous thing here? Why can't the followers see the Donald's one thousand- and one-character defects?

Love lets you see only the positive side. Donald represents power; Donald tells it like he is, racist, misogynist, me-centered. He does not feign virtues, political correctness is not for him. One follower put it well: Jesus is our savior, Donald Trump is our president. In other words, the Donald is God to these followers.

This is the thing, gurus become God to followers. The Donald's followers want to be like Donald. In their case they already are much like him, what they lack is the power. But when they follow the Donald, they feel empowered. Donald will save them.

As similarly, the most limiting line of Christianity is, Jesus saves. Jesus embodied love; if you cultivate God's virtue of love following Jesus, you too will be saved as Jesus was. But Jesus saves? Anyways? This easily becomes a popular dogma because of the perpetual human weakness—we cannot refuse free lunch.

Dogma is also a domination thing; the religion's selling point. Jesus saves only Christians.

In the case of scientists of materialism, they, too, have gurus, Newton, Einstein, in that ilk, the founders of objective science. Objectivity is to be saved even when we are studying consciousness which is about both subjects and objects. If all is matter, if all is mathematical, then objectivity is never under question.

Look at our post-materialist, followers of holism, systems theory, what have you? But all about objects. Science has to be objective, said the great David Bohm. He was a friend of the spiritual wisdom teacher J. Krishnamurthy. In their jointly written book, *The End of Time*, Krishnamurthy is telling him about nirvikalpa samadhi, direct experience of reality but subjective, and Bohm struggles and struggles but doesn't get it.

But objectivity is a great way to subdue religion, even subdue idealism. Domination.

We know this is a highly simplistic rationale of the dogmas of thought systems. But this is one way to look at divisiveness if we want to integrate and include. There is a positive side to everything, even criticism.

What is interesting is as all things get said and done, quantum science gives us some objectivity back. The physicist Bernard d'Espagnat calls this weak objectivity. Strong objectivity of the materialist demands that data is observer independent; for weak objectivity, we demand the data to be observer-invariant. So long as the data does not change in essence from observer to observer, we are ok. The data is replicable; theories are verifiable.

God and the Guru

People often ask us, Is there any role that religion can play in the future as science and spirituality are integrated into one single system of wisdom? There is. This is why the need to look at them from the positive side. Looked upon this way, every religion is defining a path, the guru's path. Although, ultimate truth can never be found by following a path because it requires a quantum leap involving looking at the archetype in suchness. But the guru can help set an archetypal context within which there is scope for creativity as well—situational creativity.

Practicing the Presence of God in Your Life

We now can tell you what the big push is for reviving God in our personal life and even in our social conversation. Yes, revival of God and moral values cures religions and conservative thinkers from perpetually feeling that they are being attacked with annihilation in mind by the opposition. It also opens the door for religions to relax some of its unnecessary dogma, for example the opposition of the concept of evolution. Some work has already begun in that direction. Read the book *Thank God for Evolution*.

Why thank God for evolution? You know what puzzles many people? When they see evil, they wonder, Why does a good God permit so much evil in human being? Evolution gives a very satisfactory answer. Evolution, though it bears the mark of God's signatures—quantum leaps of creativity and purpose—is opportunistic, all creative processes are. Survival needed demanded that we humans developed the negative emotional brain circuits, the main source of evil in us.

This is one reason. There is another equally compelling reason. The concept of God gives us a personal relationship with reality. This is because of the nature of the quantum self.

Quantum self is like a precipice for most of us. It is the junction between the manifest and the unmanifest; but the illusion of a separate self stubbornly stays even through the exploration of wholeness. Only with the self-realization experience, and only if you enter it with proper preparation (as in Buddhism, nonattachment to the manifest world) do you see it as just a doorway to unity—as no-self. Until then, quantum self is perceived by us as a self, only more expanded in the scope of its awareness and inclusivity. This is the picture of God that gives us a very viable path to spirituality—the love path—which Indians call *Bhakti Yoga*.

Jesus himself stayed on the manifest side of reality most of his life and looked on God on the other side as "my father" with whom I in my quantum self "am one." So, he knew. Yet why did he teach the love path? The same with Sri Krishna, St. Francis of Assisi, and Mira Bai. They all knew the no-self option of the quantum self but stuck to singing the glory of bhakti. Why? Because, the love path makes it all so easy. Love is an archetype in which the feeling representation dominates the thinking representation. Feeling is uncorruptible, not susceptible to conditioning unless you mix it with either sensing or thinking. This is the key to Bhakti yoga.

The authors know. We each have a story to tell. I will tell my story here, from the male side. Valentina will tell her story in chapter 6, from the woman side.

Amit's Story

I began my spiritual journey with a big handicap. The truth, I was an intellectual, a materialist at that. The saving Grace was that I was not a successful rationalist like the Average white North American or European male, I did not suppress emotions. Like most Indians growing up in Indian culture, I expressed. What that means is that I did not deny that emotions exist.

So, when my wife (she was white and North American, but being female also expressed emotions) challenged me to "love from my heart" initially I was at a loss. What is heart? Emotions exist, but are they not brain phenomena like all experiences?

Well, this was seventies America. There were plenty of teachers from the East who taught about Bhakti. There was Swami Muktananda who could touch you with a feather and you would feel prana, vital energy.

My initiation to vital energy came in stages:
First stage. In those days, I was also hobnobbing with psychologists, my colleagues in the Psych department of the university. One day I

received a call, an urgent request: "Amit, there is a fellow here who claims he can help us feel nonphysical energy. We are not hard scientists. Please help."

I went and met the fellow. He looked like a hippy, had a pretty girl with him, and he told me that he could demonstrate that we have a nonphysical energy body.

"How?" said I.

"Oh, I rub my two palms like this; they become energized." He rubbed and held his hands making a gap between them.

"Now you put your hand in between my energized hands, and you will feel a tingle. Why the tingle? You are not touching anything. The tingle signifies that you are experiencing my nonphysical energy body."

One by one, the psych professors put their hands through the fellow's hands. "You feel anything?" I asked. One by one, my colleagues shook their heads. "No."

Finally, my turn came. When I put my hand in the gap between the fellow's hands, I really did not expect anything after so many people came up empty. But surprise, surprise, I did feel a tingle, an intense tingle. When I told the psych professors, it was their turn to be surprised. "Are you sure you are not being naïve?" One of them blurted out. The others did not say anything, but their looks indicated that I had lost my credibility with them. When I looked at the hippy fellow, he winked.

Second stage. In 1981, I was at the Esalen Institute in Big Sur, California, as a guest lecturer for a week. The late spiritual teacher Osho, Bhagwan Shri Rajneesh at the time, had a big following in America then. I was invited for a morning meditation with a Rajneesh group and so I went. Somebody explained: the meditation would consist of four parts. The first was just shaking your body standing in place; I discovered it really wakes you up. To begin the second stage, somebody shouted stop; the instruction was that we stop right where we were and meditate in that standing position for a while. That passed. At the third stage, we began slow dancing to music with eyes closed. I was doing great, but suddenly I bumped into somebody and opened my eyes. And lo! I saw a pair of bouncing boobs before me! Oh, did I mention that the Esalen Institute was quite famous then for allowing a lot of nudity? Also, did I mention that in spite of being in the USA for some time, I was still kind of unused to nudity? So my body reacted in that peculiar protrusion that a man's body is capable of and I was thoroughly embarrassed. Fortunately, the bell rang and that was the signal for the fourth stage: we were supposed to sit down and meditate, which I did. But the feeling of embarrassment persisted, and right then a strong feeling of energy rose from my anus to about my throat, maybe beyond. And it was quite delightful.

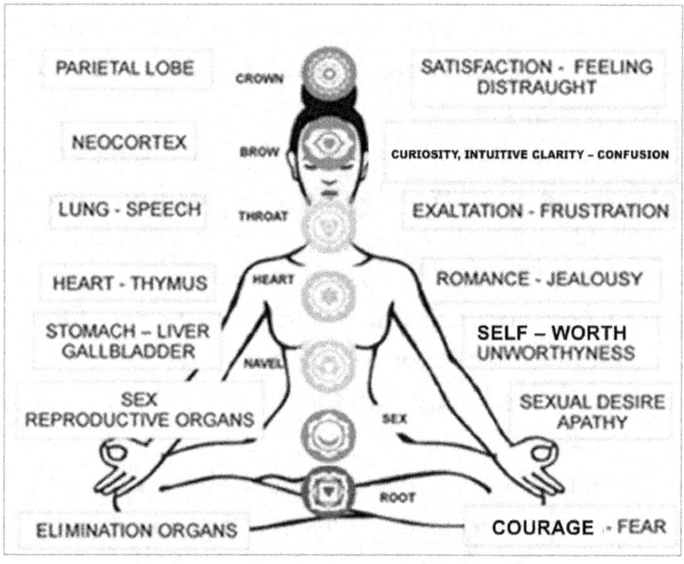

Figure 2 The Chakras: the feelings are shown at each of the chakras when the heart and navel chakra awaken.

Now mind you, I grew up in India and in that culture, everybody knows about the chakras; chakras are the centers where we experience feelings; they are located approximately along the spine continuing into the brain. Fig. 2 shows you the different chakras and the feelings *psychologically healthy* people experience at each of the chakras.

Note: For post-materialists, there is a major misconception about feelings that we experience in terms of a misleading "vibrational model" for the origin of feelings—the idea that different feelings that we experience are different because of the difference of the frequency of the vibration of the same basic stuff that is producing the feelings. In this way, it is proposed that the experience of the feeling of security at the root chakra morphs into the feeling of love in the heart chakra because that basic stuff is now vibrating at a higher frequency. This model has no science behind it simply because there is no such basic stuff in the body; instead, what we have are different organs performing different functions.

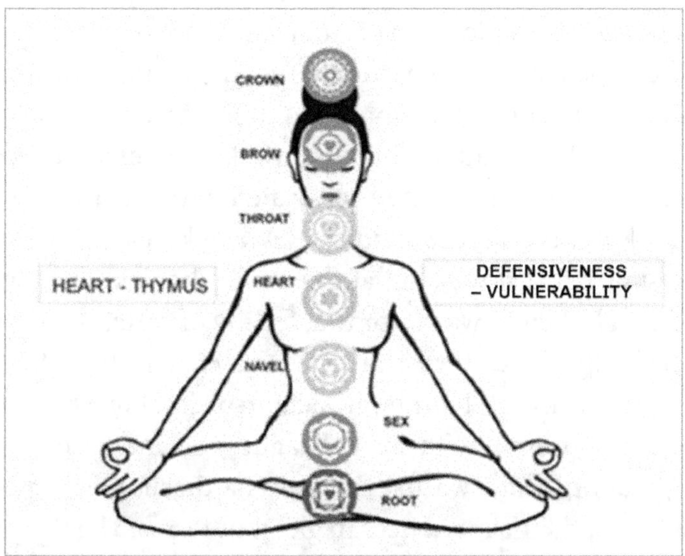

Figure 3 Feelings at the heart chakra for most people—defensiveness and vulnerability respectively.

In chakra psychology validated by quantum science, the feelings at different chakras are different because the organ functions there are different. At the root chakra, the elimination organs provide us a crucial component of nourishment; naturally this is associated with the feeling of security. At the heart chakra on the other hand, usually the immune system whose function is to defend us from intruders dominates giving people the feeling of defensiveness, a signature of sub-normal personality (fig. 3). Mentally healthy people are able to suspend their defensiveness and become vulnerable when they interact with a suitable partner; it is then that the heart becomes coherent and quantum and the feeling we experience at the chakra is love.

Kundalini awakening is an experience in which you supposedly experience movement of *prana* (vital energy) from the lowest chakra (the root) to the highest one (the crown). The Sanskrit word kundalini means coiled up energy. The idea, expressed in quantum terminology, is that the feelings at these chakras partly remain in potentiality until a sudden quantum leap of awakening takes place when they open some more, some of their previously unmanifest potentialities become available for manifestation. So, at once I romanticized that I had a kundalini awakening experience; but I was also disappointed that the energy did not rise all the way to the crown chakra as the literature indicates. Was it kundalini awakening? I am now sure that it was, but at the time I didn't know. I was not aware enough then to watch for that feeling of certainty in my gut!

Third stage. The third time was the charm. In 1983, I was attending a workshop by the physician/spiritual teacher Richard Moss. Richard (along with his colleagues) gave each one of us a "chakra healing." Afterwards, there was a discussion, and many were telling of their experiences in glorious terms. The twenty-five people I shared the session with seemed to have some sort of energy exchange, while I had hardly any experience, and felt totally left out. Like "What am I doing here?" Finally, I could not hold it any longer and raised my arm.

"Yes, Amit."

"Richard, looks like you gave all these experiences to all these people, why not me? "

Richard said, "Amit, I can only open the door for you. It is you who has to pass through it."

"Sounds well and dandy. So, you are saying all these people passed through the door and I chose not to?"

"That's for you to decide. All these people left their selves at the door. That's the trick. Then you enter."

"But I am a scientist. I want to be there when it happens." I blurted out. Everybody was laughing (or LOL in today's language). And I realized my mistake. In the next few days I received ample doses of Richard's prescription medicine—juicy physicality consisting of intimate hugs, especially from women. After that and a few more sessions of chakra healing, I got it. I could experience energy in my chakras.

Fast forward. I learned to love my wife from the heart. Not only that; exploring quantum physics led me to a big breakthrough which romantics would call gnyana yoga realization of God. But to me it felt like a wonderful creative experience, I had solved the quantum measurement problem with the great insight that consciousness, not matter, is the ground of all being. Quantum leap, yes! A big one? Yes. But self-realization? No. God realization? A resounding no.

I made progress in understanding and putting my understanding into practice as far as love is concerned though. I discovered a scientific theory of feeling, the chakras, kundalini, the whole shebang. I even practiced a little of what I called Jesus' yoga. Jesus said, "When the male shall not be male, and the female not be female, then shall you enter [the kingdom of God]. To me, this meant integrating the navel and the heart chakra. This idea is now part of quantum yoga. *The first, navel, is where we experience self-worth, and even self-love; the second, heart, is where we experience other love.* I took the first step; I came to terms with a complementarity relationship with my wife. She would help me expanding on other love; I would help her find her way back to self-love.

New circumstances arose and this was not enough. I intuited that I have to take the next step—learn to love unconditionally. How does one do that except by totally integrating the navel and the heart? I began practices to that end. I made progress. Quantum science seems to predict that the integration would lead to the two chakras fuse together at the higher functional level and I would have a big experience of opening of the heart. Something like what Jesus did to attain the heart of Jesus.

Years ago, I had a spiritual teacher under whose prodding I read a book by a great Christian mystic Brother Lawrence, *Practicing the Presence of God.* I even tried to practice it for a while; not much success at the time. I really did not know what it meant. Brother Lawrence was miserly with words; did not explain much.

Another intuition came: why don't I *change* the practice slightly to Practicing the presence of Love in the heart? Didn't Jesus say, Whoso knoweth love, knoweth God? That should be doable. It was. I did it and did it and did it, like *japa* in which you take a mantra and repeat it silently. The mantra internalizes, this stage is called ajapa japa. Japa without japa. The japa just seems to go on; even when you are not actively doing it, as when you do other chores. So interesting.

So, then it happened. I discovered and experienced pure feelings in the heart, no thinking, no involvement of the brain. There was a gap before thinking came. Was it the feeling of unconditional love? Subsequent experiences keep confirming the truth of the phenomenon.

But frankly, as a teacher I must admit that the way I did it takes a lot of perseverance and commitment to consciousness research. Is there an easy way?

There is.

Shiva *Linga*

Linga is a Sanskrit word meaning penus and yoni is the Sanskrit word for vagina. Hindus worship linga penetrating yoni as a symbol of God penetrating Mother nature to produce creation. In my youth my Moslem friends used to make fun of the imagery.

As in the following teaser from one of these friends. God and Allah were having a battle, who is superior? Of course, Allah is. Didn't Moslems defeat Hindus time and time again in Indian history? So, at a fierce moment, Allah chopped off Shiva's linga. Ever since, Hindus worship it. You know what Shiva did? He said to Allah, ok, now that I don't have a linga, what do I do with all this pubic hair? He ripped off all his pubic hair and put it on Allah's face. Ever since all devout Moslems carry a beard.

Much later, with quantum science to guide me, the symbology of lingam and yoni took me to an active imagination that immediately aroused the spiritual heart. Go figure. This I can teach and this one leads to an opening of the spiritual heart and presence of God's love in a short time.

You may think that this practice is appropriate only for the female of the species. Not so. Comprehend the meaning of this true story about Mira Bai, one of the great spiritual woman saints in Medieval India. Mira went to *Brindaban*, the birthplace of Krishna, very popular with Bhakti practitioners. Mira wanted to be a disciple of the greatest teacher in town naturally, Jiv Goswami. But Jiv refused her. "I don't take female disciples." To which Mira wrote back, "In Brindaban, isn't every seeker a female?" The reply so impressed the Swami that he changed his practice and started accepting women as disciples and of course, Mira was his first and foremost.

In this way, the practice is for women and men. You are taking advantage of the fact that your brain cannot tell between imagination and reality. Your vivid imagery will give the message to the sex chakra and create a lot of energy there. Make an intention to raise all that energy to your heart chakra. You can also use yogic mudras and bandhas designed for raising energy at this point. Now you relax. Then active imagination again. Relax again. Do-be-do-be-do. It will work.

In the seventies, I really liked a Playboy cartoon. A guy and his girlfriend are in bed together. The girl friend, naked and all, is clearly amorous. But the guy is not interested. "Cut it out," he hollers. "Don't you see I am reading?" He is reading *the Joy of sex*, a popular sex manual at the time.

Later I wrote in a book, *The Visionary Window*, citing this cartoon admonishing my readers, like this guy you would rather read, rather want to love God than loving God who is ready to make love to you.

Now I know better. It takes hardly any vital energy to read pornography and masturbate; but it is a lot of work to make love with a partner. Motivation to change is all about energy. We have been teaching it the ineffective way all this time. *Gyana* before Bhakti. That's the way I did it, too. Took me years.

Quantum science is suggesting a better way. Do all this together. Quantum Yoga is Integral yoga employing all paths simultaneously.

CHAPTER 4

What happens after Death?

Let's talk about the problem of experiencing enchanted reality from the religious side. Religions and their followers traditionally and continuing through today believe in dualism: a world of God, heaven, separate from the world of humans, earth. The priests control their parishioners mainly through fear of hell. But where does hell fit in in this two-world description?

Science and dualism are not compatible; the problem is interaction. Two worlds that have nothing in common cannot interact. But of course, reality does not have to be scientific. In this way, religions have successfully preached to their parishioners for millennia.

From modernist point of view, the concept of hell seems like a joke. The jokes started as early as the times of St. Augustine. Augustine, you know, great speaker, he gave inspiring sermons about God, heaven, and earth. But of course, everywhere there are wise guys, backbenchers, who have a critical mind. One day, after the sermon, one of these guys spoke up. "Hey, Augustine. You are always talking bout how God created heaven and earth. Pray tell, what was God doing before he created heaven and earth?"

Good question, huh? Even St. Augustine was quiet, but only for a moment. Then he came up with a quip: "God was creating hell for people like you who ask such questions?"

When you look at the chaos today that the worldview polarization between scientific and religious dogma has created, you got to wonder. Evil is like addiction; people cannot keep away from evil without some restraint of punishment. Not that fear of punishment stops everyone, but it has been a major damper against evil in the world in the past. So, the question, now that we have a monistic view of God, heaven and earth, is there any serious place for hell? Amazingly, there is.

A quick recap. God is Oneness. As God manifests and the oneness consciousness, splits into a subject and an object first through the mechanism of tangled hierarchy in the brain and second, through dependent co-arising of reflection in the mirror of memory producing an almost complete forgetfulness of the God-origin in our self-experience of the ego.

With this picture, let's ask, where do we go when we are not consciously experiencing the separateness of the ego and the world of objects? Easy. We go back to oneness in the form of the unconscious. That is what sleep is about.

Right. So, are we with God every night when we are in deep sleep? One wishes. If this was the case, we would be able to wake up every morning and say, "I slept happily." How else would it be? You just spend 6-8 hours in Oneness, in heaven of God's company! But of course, sleep is not always peaceful. Why not?

When we sleep, we are in the unconscious; we are one with the unconscious. This does not mean that movement stops; even in the unconscious, possibilities are processed with you as the processor. And here is the catch. If you have day residue, stuff that you did not finish during the day, you will continue to process those. This is called unconscious processing. Another way, unconscious processing can create hell for us is that we have stuff of unhappiness that we suppress; and they try to come up too. This produced emotionally upsetting dreams and restlessness in deep sleep as well.

Now let's ask, where do we go after we die? I once was reading a book named *The American Book of the Dead*. In the very first page, the book said, "How do you know that you are not dead right now?" For a moment, a shiver of fear went through my body, "Indeed, how do I know?" Then only I realized my mistake. For a moment, I was thinking like Hollywood movies project. After death, we live exactly like on the Earth side. It makes no difference that we do not have a physical body anymore. We look the same, talk the same, etc.

What does quantum science say? Well, part of our assumption is right. We are fundamentally Oneness manifest; Oneness—consciousness—never dies. It is the ground of all being; where would it go? After death, we go back to the potentiality of Oneness; but we are one with it? It depends on the possibilities we process. If we are able to process new possibilities—the archetypes for example, we are in heaven, no separateness.

Death is like deep sleep, except perhaps a much longer repose. And then again the question, is the rest peaceful? If it is not interrupted by occasional waking up, it certainly makes sense that all those repressed and suppressed memory of our personal unconscious will surface for unconscious processing. And who will be processing those?

We won't go into details (which you can read in Amit's book *Physics of the Soul)* but what survives your death is what Indian savants of Yoga Psychology named *karma* millennia ago, the propensities—habit patterns and character traits—that you have developed till death. The character traits are you; they are no good for precipitating experience, for that you need a physical body. But you as your character traits can engage heavenly unconscious processing.

Now think, what it would be like to be with all that pain and trauma that you suppressed throughout your life, especially in childhood. It would be like hell. Well, it is hell; it is what religions call hell.

There is a book of the dead called *The Tibetan Book of the Dead* written millennia ago which is not written with the idea of best-selling gimmick but for giving real spiritual guidance. The Tibetan Book of the Dead says that after death, you (this character) will go through three bardos or passageways. The first bardo after death is called clear Light; the clear light of Oneness. If you are qualified to see it, you disappear into God. The Easterners call this liberation or Nirvana. The second bardo after death is called dim Light; this is a representation of Oneness sort of, a part of our unconscious that is beyond the personal and consists of humankind's collective memory. The psychologist Carl Jung discovered it and named it collective unconsciousness. A few people recognize it.

Most people inexorably then proceed to the third bardo, the bardo of rebirth. On the way, you process your personal unconsciousness—the stuff of your personal hell. Most notable in this processing is the hungry ghost realm.

Of course, the big question is ok, this is a picture that the Tibetan book of the dead depicted and quantum science explained some millennia later, but where is the empirical proof? And here is where also we have made enormous strides. The first bardo after death—the clear light of Oneness—has been verified in what people call Near-death experiences. People die from cardiac arrest, and then cardiac surgeons revive them, and they tell their experiences; that is NDE. The veracity is highly documented and quantum science can explain NDE in all its detail the most important component of which is the idea of delayed choice—the experience happens retroactively starting at the moment of your brain's revival. Read Amit's book *The Everything Answer Book* for details.

Amazingly, at the other end, the psychiatrist Stan Grof, via a process called holotropic breathing, has been able to regress people to their birthing experiences including what can be called near birth experience. The quantum science explanation again is delayed choice, similar to that of NDE.

The near-birth experiences that people reveal talks about heavenly realms and hellish realms, the good stuff of our personal unconscious and the unhappy stuff.

In this way, the heaven and hell that people visit are their own personal creation—memory of good deeds and bad deeds. Raises the question: what defines good versus bad? It is easy. Good is that takes you toward oneness; bad or evil is that takes you to separateness. In oneness, you include others and engage higher needs; in separateness you are me-centered and engage those negative emotions, pleasure with addiction, and trivial pursuits.

You know, spiritual societies have recognized this in terms of mythical stories. Here is a story from the Thai tradition. Heaven and hell are both big banquets that you attend. You sit in a same table, are offered the same food, except that you bring your own mind set that makes the processing heaven or hell. The eating utensils—the forks, knives, spoons—are all table length. In hell, people try to feed themselves to no avail and experience the hell of frustration and anger. In heaven they happily feed the person sitting opposite.

Going back to the after-death story, you not only survive after death but are reborn. One of the triumphs of post-materialist science is the vast amount of data that has verified the idea of reincarnation and quantum science explains the data in all details.

The reincarnation of individual ego/character gives a human being many more that one chance to arrive at his or her life's purpose. There is infinite quantum potentiality for each human being to explore; unfortunately, this needs creativity, and we have considerable resistance against creativity. The tendency to maintain status quo is called *tamas* in Sanskrit—mental inertia. It takes a few reincarnations and a lot of suffering to want to change.

Initially, the kind of change we desire is short term—the immediate relief of our suffering. The creativity needed is situational creativity. The corresponding quality of mind is called *rajas* in Sanskrit. The sound of the word rajas should remind you of the word raja meaning king. This is because kings and the like in the olden days had plenty of this quality. You invent some creative way to win battles and then use the same basic method with situational adjustments to build an empire. The kings were also guided by wise ministers; often these ministers provided them with the archetypal context of the rajasic activities.

Today's people of power and wealth do the same thing as the olden day kings and aristocrats. They too sometimes use gurus and guides to provide them with the archetypal context for their creative actions. But not always.

After some reincarnations with rajas-ic exploration, the ego/character becomes adventurous and wants to explore long term problems for which they need to address and explore the archetypes directly. This is when the going gets tough. The creative quality needed is called *sattva* in Sanskrit. And you have to do it alone, chart your own path; others can at most collaborate with you in your journey.

Yoga psychology refers to these three qualities of the mind—tamas—inertia, rajas—propensity for situational creativity, and sattva—propensity for fundamental archetypal creativity, collectively by the Sanskrit word *guna*.

In this way a reincarnationally mature ego/character, often called the soul, has a mixture of all three gunas. And then only, people qualify to select an archetype when they die for exploring in the next incarnation. This, your own chosen archetype, is called *dharma* in Sanskrit. Please note that dharma is spelled withy lower-case d to distinguish it from Dharma—the absolute Oneness—spelled with capital D.

You may have noticed. When you do things that you choose, accomplishment gives us more satisfaction than when somebody else is dictating what you do. When you explore your own dharma, it brings bliss. This is what the mythologist Joseph Campbell was referring to when he would urge people to "follow your bliss."

Let's go back to karma, those propensities carried with you as nonlocal memory but not part of who you are. So, you are not born with them; they have to be triggered by your life's circumstances as you grow up.

First, there is the good karma. Their purpose is to help you carry out your desired exploration of your chosen dharma; you chose them too. A good way to find out your dharma is to introspect on your propensities that when used bring you most joy. Another way is to analyze your dreams in the way of gestalt therapy. You literally sit on a "hot seat" and try to narrate your dream with all its finesse and emotional tones to a skilled guide who helps you analyze the dream. Dream objects represent your various aspects except those few that represent your archetype.

This then is the nutshell of how your life propagates through manifest time and space. In each incarnation, the higher purpose of life is soul-making, how to make your ego-character rich with gunas and other traits that make you happy, healthy, prosperous, and intelligent.

A note in passing. If you are wondering why we bring bad karma with us as well—those propensities of nonlocal memory that create obstructions to our desired journey. In addition to your drive toward wholeness and Oneness, there is also a drive toward separateness. The base-level ego tries to perpetuate itself and sabotages your drive toward satisfying higher needs. This is what gives you masks of personalities. Why do you bring them with you? To disassociate with them and become more authentic.

A lot of people take rebirth in the same family and repeatedly contribute to a "family secret" such as child abuse. Today, we have discovered family constellation therapy; skilled therapist work with such people and make them "see" the situation. With awareness, the bad karma just falls away. Read *Quantum Activation: Changing Barriers into Opportunities* that Amit wrote with film-maker Carl Blake and family constellation therapist Gary Stuart.

CHAPTER 5

How May You Empower Yourself to Live Enchanted Reality?

The Quantum Secrets of Soul-Making

Religions defined by a particular spiritual path and not by dogma is something for the future. In the past religions have been elitist—power grabbing by a few who interprets God to ordinary people of base-level human condition. True, no religion has ever demeaned us as much as materialist science, no religion has ever relegated us to robots with experience, but religions, too, have robbed us from our personal power, our prerogative, our responsibility.

We gave one example in the previous chapter, the dictum *Jesus saves*. Other ways of taking your power away is to assert, surrender your will to God's will. Or to declare, it is all God's will. Still another one happened in India, manifest world is all illusion.

Materialist scientists have replaced God by the concept of material mechanical laws. The laws govern you. But the idea is the same: to make you powerless, so the elites can manipulate and use you.

Either way, religious dogma or scientists' dogma: the idea is to convince you that you are a nincompoop. With all the appearance of doing so much, some people work for 16 hours a day, today you do nothing except make a living to entertain yourself what the talented and gifted dish out.

And mind you, the message the elitists give is not exactly wrong. What the elites are not telling you is that this is only part of the story.

Now listen to what wisdom traditions say to set your perspective right. In the Bhagavad Gita, Sri Krishna, the wisdom teacher says to his disciple Arjuna, Action that follows action are non-action. Only action that follows nonaction is action. Go figure.

Comphrendez, Understand? Seems like a riddle at first. The meaning is this: conditioned action that follows from other conditioned action is nonaction, keeps the world going in circles, nothing really changes. Democrats or Republicans, what's the difference. Sh-t from either side!

Okay, the last statement is an exaggeration. Today, many Republicans would rather have powerful dictatorship than democracy. That is not just non action, that is evil action. (By the way, the Sanskrit word for nonaction *akarma*, which also stands for evil action.) Similarly, many democrats want socialism which also compromises democracy, though not as much as dictatorship.

The other part of Krishna's great saying is the key to understanding transformation. Real action follows from nonaction. Nonaction, not doing but being, takes us to our unconscious Oneness, the source of our creativity, creative action.

If you want to partake creative action, you must learn to change your do-do-do style of living with do-be-do-be-do lifestyle.

In my (Amit's) own life, for most of it, I was into outer creativity much more than I was into inner creativity. After my great insight—consciousness solution of quantum measurement problem—I was invited to this big conference at a Yoga Institute in India. I gave my spill, it went well; people were appreciative, even the spiritual head of the ashram wanted to see me. You know, I went with a swelled head. Imagine my deflation when the fellow's first question to me was, "Professor, what do you do when you are by yourself?"

What did I do? I fidgeted, looking for something to do, a research problem to think about. I was creative of course; outer creativity can keep you busy and once you get the hang of it, you don't need to relax—be—as much. But inner creativity is a different beast altogether. You really have to incorporate do-be-do-be-do in your lifestyle. When I did, things began to happen inside me.

In this way, Krishna's advice for action following from unconscious processing is corroborated by the quantum science of creativity. Much data by field researchers in its favor. It works. Do-be-do-be-do for a while engaging with your problem for a while alternating between active imagination and relaxation, suddenly insight comes, aha! With a surprise giving away its secret, the insight is the result of a *discontinuous quantum leap of thought.*

The wisdom traditions intuited the basics, quantum science is filling in the gaps to give you a methodical way to approach transformation. To take back your power. To elevate you from the base-level human condition.

Most of humanity is not ready. They are mesmerized by the elites. They really believe that the leaders will do it for them. NEVER MIND THAT THE LEADERS ARE ELITISTS; THEIR PRIORITY IS TO MAINTAIN ELITISM. They want you to live in fantasyland, one WAY or another.

The psychologist David Hawking figures based on kinesiology—muscle testing, muscles lose their vitality when we lie—that there is about 15% of humanity who are ready to break the hypnosis, break away from the shackles of both the incomplete dogmatic worldviews of today. If you are reading this book, you are likely to be one of this fifteen percent.

How did you break away from mass hypnosis? It is built into the quantum science within the primacy of consciousness, a phenomenon called reincarnation. Finding your power to live reality requires many incarnations; reincarnation allows you the time necessary.

You see, your life beyond the base-level condition is about soul-making. The romantic poet John Keats of another time saw it correctly. He wrote to a friend in exuberance:

See the world
As a vale for soul-making.

Later Carl Jung said the same thing: purpose of being human is to search for the soul. How we make the soul is by embodying the archetypes as best as we can. The material hardware we have is not able to make direct memory of the archetypes; we are only able to represent our archetypal insights with our mind and the vital. *When we make a quantum leap of insight to an archetype, our mind is elevated to think new meaning; at the same time our organs, especially the organs at the heart and navel chakras get new software and reveal new functions: the heart wakes up to other love and the navel to self-worth. The new vital and mental software the manifestation of these feelings and meanings create is what is traditionally called the soul.* The Sanskrit word for soul in the current context does not exist and we need to make up one.

The knowledge of the archetypes in suchness is all perfect knowledge for which the Sanskrit word is *bodhi*. One who has awakened bodhi is called *buddha* in Sanskrit. Of course, the problem is that the word buddha is already taken; it is virtually synonymous with Gautama—the father of Buddhism—Buddha with capital B denote the enlightened Gautama.

So, for the awakening of the soul in the present context, we will use the word buddha with lower case b. This is important; name matters. Realize this: by awakening to the soul level of being you are becoming little buddhas. The end of the soul's journey is the Sanskrit word *bodhisattva*. One who has achieved wholeness and waits at the door of Buddhahood (perfect enlightenment with awakening of no-self) until all beings are ready to cross that threshold.

We need more than one incarnation to make soul and take it to fruition; that is one reason for building reincarnation in the schema of human science. The other reason is whenever civilizations decline, who rescues us but "old souls" who do not succumb to the mass hypnosis?

Reincarnation and You

Please recall! What survives and transmigrates after death from one incarnation to another is nonlocal memory—our character traits, habit patters, and their associated repertoire. Collectively, the nonlocal memory, the habit patterns and character traits transmigrated is *karma* or *samskaras*.

Understand clearly! The character is our essence—who we are. This is what we are built to honor and others too. Mass hypnosis caused by the worldview polarization has undermined the concepts of character and honor, have you noticed?

Character comes with not only honor but also conviction. When a habit pattern is integrated with our being, the habit—which we could use or not use—becomes a defining trait for us, our being then appropriate doing flows automatically when the occasion calls for it. Honoring your character is honoring yourself. It is tantamount to self-respect.

Karmic habits can be divided into two categories—good and bad. Good karma helps us; bad karma creates obstacles, they have been the side effects of the "sins" we commit, the actions that we take that increase our sense of separateness. In each incarnation, we bring a certain amount of bad karma to get rid of it. We bring good karma to help us with soul-making.

On closer study, researchers find that we bring only a portion of the good karma, too. Why? Because we limit our soul-building to a manageable number—one or two or three at the most, of the archetypes. The chosen archetype(s) defines our *dharma*—individual purpose of this life.

Then in the current life, we use our character traits and good karma to explore and embody the archetype in the process also converting the good habits into character traits with conviction. That is how we create a new "me."

The old "me" of the base-level condition is defined by its specific conditioning and the masks of personality. The new me has additionally, a character, something to honor and cherish. The implicit "I" is a little more explicit for the new me, the soul level of being.

There is one more aspect of the journey of reincarnation from one life to another. This is the concept of the *guna*—a Sanskrit word to denote qualities of how we use the mind for creativity.

First, we can leave the mind as is with its existing repertoire of meaning that we can think as available mental information. This quality, much like the material quality of inertia, is *Tamas*.

The second quality is *rajas*. Rajas is the kingly quality of empire building. Like the king, once he discovers how to invade another land and grab it, wants to do the same thing over and over, so does the human with this quality of creativity. This kind of limited creativity is called situational creativity.

The best quality is sattva, the quality that enables the creative to explore archetypes directly, to explore the new archetypal context needed for every situation whereas people of the quality of rajas are happy to work within the same archetypal context over and over.

Little that we know of the spiritual history of humanity, it seems that the archetypes can be classified in the order that people traditionally have explored them. The exploration begins with the archetype of abundance, initially explored as mere money-making through trade requiring very little creativity—rajas. As the next level of human development, the archetype shifts to power requiring more rajas. Beauty is another beginner's archetype, initially used for entertainment.

Further development leads to the exploration of the archetypes of goodness and love, and these require not only rajas but also a modicum of sattva. Then comes the exploration of abstract archetypes—truth and justice. More sattva is required, and less rajas. The final two to explore for very mature souls are the archetypes of wholeness and self, both requiring a lot of sattva.

For further perspective, read our book, *The Return of the Archetypes.*

Motivation: What's Your Archetype?

When we research motivation for the people who enter the transformative path, the overwhelmingly large majority enter it from the suffering side; John Keats called this *via negative*. Why?

Suffering is blocked vital energy. When the block becomes humongous, we cannot bear the suffering anymore and begin sincere intentions for healing then only we become sensitive to intuitions and archetypes.

But if a person is even borderline happy, has an image that he is happy in a life of pleasure and entertainment, he will not make the effort. Even though you tell him about all those infinite potentialities to explore and enhance happiness. Your telling him has given us the "professional" workshop attendee. Good for us in the workshop leading business, but the worldview will never change this way.

At a transpersonal psych conference, the psychologists Arnie Mindell and his wife Amy did an exercise. All of us assembled around them. Then Arnie called. "Ok, those of you who are giving a talk or leading a workshop in any capacity please stand on my left. A few of us stood as directed. Next Amy called for people who are at the conference because they have research interest. Again only a few responded. Next Arnie again. "Those that are here to learn, check out new stuff, etc. Please stand over there." He pointed and a relatively few more people, larger than the two previous groups to be sure, moved as directed. This surprised me; I had assumed that this was the motivation for most of our attendees. Amy gave the final call. "Those of you who are here looking for relationships." To my surprise, this was by far the largest group.

Research shows that a few people who go into transformation today from the positive side have a lot of enthusiasm; the enthusiasm manifests as curiosity about knowledge for men, and as other love in relationship—service—for women. The first group has vital energy in the brow chakra; the second group in the heart chakra. How did they become this way? Reincarnation—karma—could be the major reason.

Even taken together, people in via negativa and via positiva today, make up a small fraction of the 15% of the people who we know are ready because they are dabbling with the idea of wanting to want God. If we want to change the worldview now that a third integrative one is available, we must energize these people to want God, to want transformation. What is the problem?

Have you wondered how human professions originated? The earliest professionals were probably artists and musicians. Cave painting, the preponderance of rhythm in tribal cultures, both suggest that the archetype of beauty was the prime motivator here.

Next came the question of security to serve the survival instinct, of course. This called forth the archetype of abundance. Next perhaps came the archetype of power and perhaps soon to follow was the archetype of goodness to put a check on power. These archetypes gave us the profession of politics and religion. Even hunter-gatherer societies had all these professionals.

When the anthropological era changed to garden agriculture and tribes started to live in families within the still nonlocal connected community, partners began to explore the archetype of love. Families also began to trade, barter economy came about, and the archetype of abundance motivated the business profession. After a while, the exploration of archetype of goodness also helped to make the pursuit of abundance partly transpersonal.

The profession of technology (fire, metallics) must have developed pretty early as well as was the profession of the healer and judge. However, their association with archetypes might have to wait till the development of sophisticated concepts in the agricultural era which began the current rational era of humanity. Technology became part of science exploring the archetype of truth; Healing became the exploration of the archetype of wholeness; judges became part of the law profession exploring the archetype of justice.

The last archetype to be discovered was the archetype of self, circa 5000 BCE. Finally, the tradition developed that as your archetypal exploration arrived at some mature place, you had the option to teach—pass on your wisdom to others. Teaching profession was created.

Even today, these are the professions that dominate human society. Unfortunately, the professions no longer offer you the opportunity to explore an archetype necessarily. How the society looks upon the professions has changed.

The Job Mindset

The rulers of the British empires had colonies with established old civilizations, the prime such colony was India. Since they were the conquerors, they thought of themselves as superior to the colonized people and forced the natives to give up their professions and serve the new masters. In this way, the professions were changed into jobs that people do to survive, not to explore archetypes.

Later, when capitalism came about, industries and tech businesses discovered that the job mindset is more profitable and that became the way of most human societies. With the advent of materialist science, education, too became institutions of job training with the last vestiges of archetypal exploration in professions disappearing as the archetypes themselves were officially purged from academe.

A job gives satisfaction only to a person who is in charge—the satisfaction of accomplishment. The underlings get so tired working for somebody else all day, that after work also, they do not explore meaning and purpose questions anymore. Instead, they settle for entertainment. In this way, the entertainers, the scientific researchers, the tech innovators, the leaders excepted, the whole society becomes a wasteland with people (even the fifteen percent) denied the opportunity to explore their archetypes, ever wondering about what their archetypes are.

The Added Burden of Rationalism

On top of all this, consider the added burden if you are mental, with me-centeredness and negative emotions defining you. Life of a job holder becomes a difficult ocean to navigate. You are at risk to get manipulated by the elitists, the leaders of the society.

Today, we have two kinds of elitists and each manipulate you somewhat differently, but manipulate they all do. The meritocratic elitists (knowledge is power) cater to you only if you are meritorious enough to be a college graduate. Then they manipulate you via progressive slogans, a bright future with robots doing all the menial jobs, freeing you from everything mundane never mind that this kind of progress would leave the nonintellectuals, people of no college education, at an utter disadvantage. Think! Do you really do a non-routine job yourself? Although your job is mental now, since it is routine and fundamentally logical, sooner or later machines will take over your job too.

Meanwhile, the materialist rational culture tells you to suppress emotions, their one recipe for dealing with the preponderance of negativity of your brain. Values are denigrated, materialists do not believe in positive emotions or archetypal explorations that generate them. They do however, acknowledge that some semblance of values is needed for civilized existence. With negative emotions suppressed, the meritocratic recipe of positive psychology is thus enjoying pleasure and pretend the archetypal values for social relationships. In other words, all relationships are still transactional but with a lot of value-pretention.

The other elitists are old-fashioned aristocrats, the nuovo rich plutocrats, and the religionists. Materialist science has no credibility for this group and on the face, they support religious/spiritual values except that they do not follow the values themselves; they are hypocrites. However, the values enable this elite to manipulate and rule the non-meritorious people, the grade schoolers, by fear and divisiveness. Anyone who does not belong to your group of people cannot be trusted, cannot be included, must be hated.

In this way, most societies and cultures are now polarized between higher educated and lower educated. The new divide: "the twain shall never meet;" at least it seems that way.

Let's return to your plight now, you who belong to the fifteen percent who *could* pursue the archetypes. Please see that being a college graduate is now a trap for you. You have invested heavily, you have a huge student loan to show for it. You want something out of your college degree, a job and you are trapped. The job enervates you so much that even dabbling into meditation and yoga only keeps you functional. So, you wait till you hit a scientifically built-in mid-life transition stage of life. Men however often find another similar courier, get a new partner, and move on without using the opportunity. They rationalize.

Women see mind-life transition differently. You may have heard my favorite story of the modern middle-aged college woman. Such a woman is in a debate with a clergy man and a meritocrat scientist about when life begins. The clergyman says, "Life begins at conception." The scientists pooh-poohs the idea. "Oh come on. Life begins at birth." He smiles and looks at the woman expecting approval (abortion you know). The woman surprises both men. "Life begins at forty. That is when your kids leave home, your husband divorces you, and the dog dies."

These are the women and a few men of good karma, who are the people who are changing the world by engaging the archetypes. It is not much simpler if you are only a high school graduate and become open to the other scientifically built transitional stage. Because there is a challenge here too; to get over the job mind-set. That mind-set ties you to entertainment and fandom. Again, most men get in that trap and never bother about transformation.

For women, the entertainment world is fortunately shallower, only fashion models and stars and celebrities of the movie and TV world. If you have a weak navel chakra, most women in our current culture belonging to the lower education group do, then you associate with macho men and adapt their sports and other macho entertainment as well. You are trapped too.

But suppose you are a borderline case with your navel chakra quite strong! You will be quickly demoralized by macho men. You are now ripe for exploring the archetypes.

For women, even with a college degree in one of the "soft fields" where the man-dominated culture pushes you, it is hard for you to get a job, get a courier, and this could be your saving Grace.

In this way, we now have a substantial number of women of both groups who have broken the shackle of worldview polarization and enlivening the transformation scene. Things are changing. The development of quantum science is in response to this demand created by these women and men. It is no wonder that 75-80 percent of quantum activists are women.

What is the source of the vital energy that these women and men feel in order to continue their journey? These women have fire in their heart and the men fire in their belly from their past-life karma. If we want to expand the number of new paradigmers, we must provide the newcomers the fuel for amplifying the fire in their navel and heart chakras. Inspiration.

Belief System Clearance

The quantum worldview demands a worldview integration throwing way the chaff of both religion and materialist science. But it is not as simple as it seems intellectually. Even though today you have the benefit of experimental data along with some of the higher consciousness experiences that you have without which you would not be reading this book. You want to change; you are looking for having enough faith to catapult you in action!

My (Amit's) physics colleague, this was in the nineteen seventies, went to Nepal to teach middle school children as part of his research in physics education. One day, he taught his students how water evaporates, goes up high in the atmosphere, cools and condenses around particulates. When the condensed droplets become sufficiently heavy, gravity pulls them down and it rains. A week later, he was testing his students. A bright one spoke up and gave a perfect scientific description of rain making. After he finished, the professor was pleased and told him to sit down. "You have done good." But the student said, "But there is another way that rain happens too. When the rain god is pleased."

My colleague told me this story to illustrate how hard it is to change beliefs that have been formed in childhood because that's when we are most gullible not having developed memory sequencing and all to give us an understanding of proper perspective for meaning thinking. I am repeating his story with the same idea that he found to be true. We hold on to our old beliefs side by side with the new. This is disaster. This is how value ambiguity enters.

Today's people are value ambiguous as a whole culture. This is why value wars are becoming culture wars when people are coming from two different worldviews as their dominant belief system.

How to you clear your belief system of this equivocation? Creativity—quantum leaps—beginning with deep understanding of the new. It takes a lot of motivation. This is why most people enter transformation from *via negativa*; the suffering is a great incentive. Ambiguity creates an opening for running away from the problem! Suffering takes away that option of running away.

So, you develop an open mind, say "no" to the old, *neti* in Sanskrit, not this, ambiguity and all. Your unconscious will process it to advantage and create new possibilities for consciousness to choose from. But you have to be steadfast in maintaining the doubt the ambiguity until the belief system clears with a quantum leap.

Read Amit's book, *Quantum Creativity*, for further explication.

Changing Obstacles into Opportunities

Besides belief systems, there are many other obstacles created in your path to transformation and soul-making. Some of them are sociocultural, some your own conditioned memories, some are happenstance. You have to see that they, too, in a way, are creating opportunities for you to disengage from your usual apathy; they are giving you energy. Crisis is both danger to run from and opportunity to engage and change. One perception leads the usual fear response. The other perception requires a strong "gut," a transformed vital navel chakra with a new function of self-worth.

Predators are looking for you to find a new victim for their manipulation. You don't have to fall for their manipulation; you don't need to be a follower to feel power from borrowed gut; you can awaken your own gut. Remember! Being awake and being woke are not the same thing. The woke people are equally manipulated. The courage is real only when it is your self-worth that is propelling your action.

All this and much more is explained in details in the book *Quantum Activation: Changing obstacles into Opportunities* by Goswami, Blake, and Stuart.

The Quantum Science of Manifestation

Hope you are getting the idea: we have a come a long way since the nineteen seventies simplistic slogan, we create our own reality, and mysterious secrets of how to do it. The truth is: there is no free lunch. To manifest meaning and purpose, happiness and intelligence, the archetype of your life, in you being, you have to awaken your somewhat hidden quantum connection wherein lies your other half of your co-creating power. Much of the creative process for manifestation that we are suggesting is like old-fashioned spiritual work. But make no mistake. Previously, people were engaging it to transcend the world for which they were not even ready. The new quantum vision is to engage transformative creativity to manifest a new you so you can engage with the world happily and more intelligently until you are ready for Buddhahood with a capital B.

Quantum science of manifestation is clear in its epistemology: first you create the proper context. You have been using some version of the three R's all your life so far, job training. Time to engage with the three I's: Inspiration, Intention, and Intuition. Intuitions will bring the gale winds of change—your archetype(s).

And then, engage the creative process. Translated into I-words, it has four stages: Imagination, Incubation, Insight, and Implementation. Remember the way to employ the first two: do-be-do-be-do. There you are.

There is a story in the Upanishads. A young boy, Nachiketa, is looking for meaning and purpose of his life. When his father is doing a big sacrificial ceremony to invoke the archetypes, the boy is insistent: to whom to you sacrifice me father. The father in his irritation says: I offer you to *Yama*, the death god. But Nachiketa is not to be demoralized. He goes to Yama's abode, Yama is out of town he is told, so he waits patiently. When Yama returns and sees the boy deep in creative meditation, he is impressed. He gives the boy some boons and finally asks, what do you want to know? Reality, I want to know reality, the boy says.

Yama explains monistic idealism but ends with the words: arise, awake, and enjoy your boons first. That is the thing: heaven can wait. Find your dharma, follow your bliss, and enjoy life with happiness, intelligence, accomplishments and enchantment.

CHAPTER 6

Women's Spirituality

In recent years, the necessity of healing the dichotomy between masculine and feminine has become obvious even as an awareness point within the society. I (Valentina) saw this at all the levels of manifestation, and how an unhealthy male-female relationship at the personal level reflects as a weak, untransformed, fearful society overall. Healing this dichotomy starts with recognizing and awakening of the power of the Feminine.

A major intent of our work in quantum science is also to understand, heal, empower and raise our awareness upon genuine femininity, which includes the male-female integration within ourselves, for a deep healing and re-enchantment of our connections, locally and nonlocally, individually and globally.

We, women, need to ponder upon our feminine mysteries, privileges and qualities, as well as weak points, and emphasize on specific keys for a rapid evolution as a woman; we need to re-vision the social and cultural taboos that women are not "supposed to break" but that are absolutely necessary to reconsider in order to awaken the feminine within ourselves. It is not about becoming feminist activist in the old-fashioned way. Instead, we need to see ourselves as all that we are and learn what we need to learn and change what we need to change.

Here is a quote that I love, by Amit: *"The integration of Love, with the archetypes of Power and Abundance is so important in changing the affairs of politics and economics today. Literally, we have to feminize this male dominant fields of human endeavor. This is the major role of women in global healing in terms of science. Male dominance has produced the excess of the rational thinking and this also has to be healed. Only women in large numbers and together can heal the world by bringing the heart in the affairs of men.*››

We need to look into recent scientific advances, the quantum theory of vital energies and the chakras, as well the way consciousness identifies with the navel and heart organs as selves independent of the self of the brain, and yet, does it differently for the two sexes forming the rational basis of the male-female differences in the way people experience emotions in the body. Eventually, the idea is to understand the spiritual power and responsibility of being a Woman.

Practically speaking, we need to learn how to build a character balanced in its subtle experiences (think—feel—intuit), reaching to a balance between our relationships and our professions, and also understand and fulfil our traditional role in family. Such balancing, understanding and fulfilment take intention, commitment and inwardly directed creativity, then manifestation.

The main differences we can easily see at first glance between men and women (and yes, we need that polarity within and without in order for the species to survive, and yet integrate the polarity to access and explore higher needs in order to have Transformation in the world), manifest at 1) The brain and 2) The body.

At the brain level:

1. Women process emotions differently than men due to the way their midbrain/amygdala connects to the conscious left brain; as a result we are easily expressing our emotions (sometimes too much and then it degenerates into anxiety); for men it is the opposite: amygdala connects to the unconscious right brain and thus their capacity of suppression of emotions.

2. The neocortex is heavier in women (capacity to think is greater). The downside of this is women talk more (yes! It is true); the upside is women are better and more wholesome in forming relationships combining both thinking and feeling.

At the body level:

1. The Heart—women are more aware than men of their self at the heart being involved in the maternal and romantic qualities; women care for the other in relationship.

2. The Navel—the self here is the women's weak point, therefore dependency easily can occur unless we learn how to collapse the vital energy of self-worth at this chakra and enhance the functioning of the navel self with self-trust, self-confidence, self-respect, and self-acceptance.

Note—the woman's body presents an extraordinary complexity; when it is explored from a deeper viewpoint it reveals itself as being much more complex than a superficial glance would indicate. The woman's body proves to be more complex than the man's body due to the fact that the woman's body has certain distinct characteristics that are unique compared to the man's body.

Women can also have various types of orgasm, out of which at least two kinds are generally accepted: 1) clitoral, which is much like male orgasm, brain and pleasure-centered; 2) vaginal, often identified as a G-spot orgasm, is accompanied with happiness in the form of expansion of consciousness indicating the sublimation of the sexual energy to the heart chakra.

In contrast, the male orgasm (as well as the female clitoral orgasm), the sexual energy is felt not only at the brain but also at the navel chakra and this boosts male body self-awareness at the navel; but unfortunately, because of brain's involvement, the boosting of the navel self is entirely me-centered and enhances narcissism.

The result is again clear: the socio-cultural development of male-female difference is further augmented when sexuality awakens at puberty: males develop strong (in a narcissistic way) navel, and weak heart. Females end up with strong heart (in a needy way) and weak navel.

In this way, almost every woman tends to look for the source of her happiness somewhere outside of herself. Whether it's a boyfriend, a job, a boss, family or other desirable things - a house, a car, etc. You've probably noticed how happiness looks like something in the future, and you've often said to yourself, "How happy I will be when…."

Instead, consider what the spiritual and transpersonal psychology traditions are saying. A happy woman is one who is reconciled to herself, who accepts and loves herself as she is. When she is in contact with her soul, with the spiritual dimension of her being, the woman finds her own smooth path to travel in existence. Then everything comes to her: everything she wants, everything she needs for survival and everything she needs for fulfilment. She no longer goes outside to seek her happiness, because her happiness springs from her heart and even nourishes those around her. She shines and becomes a hotbed of feminine beneficial energy that magnetically attracts everyone else.

About FEMINISM and the Feminist Movement

After centuries of being the 'weak gender', the discovery of reliable birth-control pills and the sexual revolution of mid-sixties started the debate about women—who they are, what they represent, what their values are etc. Unfortunately, the 'feminists' of the era directed their focus upon a non-natural socio-political equalization of the sexes' as if this would ever be truly possible. The male and the female of the human species have different values and priorities; the feminist leaders forgot that.

The feminist movement took socio-political power very seriously and fought to convince every woman that she can and must become a 'kind' of a man (that is, she dresses up like a man, walks like a man, lifts weights like a man, speaks like a man, joins the military like a man, engages in playgirl philosophy like a man's obsession with playboy, and so on). The movement's promise was that after this "manning" of a woman is complete, all her problems will be solved. Surprise! after more than 60 years of feminism we have to conclude this: women are not happier, or fulfilled with how they live, or content with their jobs. Nor are they satisfied how they express their sexuality, how they engage in relationships and lifestyle or explore their purpose or destiny. They still do not know more about who they are and what is the meaning and purpose of their lives! The fact remains: the contradictions and dichotomies between sexes were not properly recognized, let alone resolved; neither did happiness, good communication or better relationships replace the old patterns of thinking what a woman is.

Instead, the feminist movement has polarized women just as materialist science has polarized our entire society. What a choice for a woman to make when the choice is between 1) a politically correct manly woman, co-opted by men in their materialist values or 2) continue as the traditional woman trying her best to hold the family and society together with her heart but always depending on the men for her power!

We are witnessing a time of huge development in the area of science and culture, new discoveries in all areas are blowing away the old ways. As the movement toward separateness continues, the informational era with its bombardment of information of all kinds of things (we know what happened in all corners of this—small world for now—within hours after it happens) has started to establish the interconnection of us all with everything and everybody on this blue planet, shallow and local as it may be. It is so easy to get excited and forget that having all this information (and misinformation, they come together, you cannot have one without the other) does not help us in making our life congruent between how to think, how to live and to make a living. Not do they help with resolving the fundamental dichotomies, in particular the male-female one, that the women especially have to come to terms with. Why this special role for women? Because women care for relationship, they are made that way.

Face it, women! The materialist information (or misinformation) culture is antagonistic to the age-old spiritual conceptualization about life and death, God, consciousness, and soul we've inherited. *If you give up* listening to the materialist rational men surreptitiously alluring with power-sharing or old-fashioned macho-men telling you what to do giving you a false temporary sense of empowerment, WHAT THEN?

First, Women need to reestablish the wonderful polarity and complementarity concerning the masculine and the feminine roles; second, they need to address male-female differences with the idea of movement toward wholeness, but in stages. They need to pass this entrance requirement before they try to attain the next stage of the journey towards wholeness.

This has been the perpetual human problem through the ages. *We make discoveries and make technologies for outer changes forgetting to make inner changes. In that way, we end up using the new technology to destroy what we are. This is what we have done from the discovery of computer power to the discovery of birth-control pills. It is time to change that habit.*

So, we find ourselves in the dawn of a new era, with new scientific conceptualizations about the fundamental matters of life, such as who we are, where are we coming from, where do we go after we leave this body with the so-called death, and what is life about—its meaning and purpose. What does quantum science have to say about what the new women of the new age should aspire for?

How to Become an Empowered Woman with a Soul Developed?

Step 1. Beginning

First direction to consider is: before being a soulful woman, one shall accept and embrace fully the condition of being woman. (Or in simple words, before becoming anything else, first she has to be a woman, that is, accept her womanhood—this is part of her destiny!) The good news here is that the quest for finding who she is—a woman—will naturally reveal her soul-potentialities.

Step 2. What is Shakti?

Embracing the condition to be a woman and wanting to do the best of it, one has to learn, also naturally, the concept of Shakti, the feminine archetype of power.

Step 3. Awakening.

The intrinsic necessity of awakening the state of Shakti for women has to be understood.

For women, developing a soul necessarily includes this huge accomplishment: the state of SHAKTI (awakened at all the levels, personal and transpersonal).

Step 4. The state of Shakti has to be manifested at all the different levels of being.

The method, laid out millennia ago by Jesus, is:

To make the two one
When the male shall not be male
and the female not be female.

In quantum science, this is understood as the creative balancing and integrating the navel (developmentally strong in male) and the heart (developmentally strong in female). This awakens in women what is called Shakti—the embodiment of the feminine archetype of power integrated with love which now morphs into empowered love. For details, read our book: *The Quantum Science of Love and Relationships.*

What is Femininity?

Femininity is a state that every woman innately has, manifested in different shades and intensities. Some women are more aware of their femininity, others are not. The more we are aware of our femininity, the more it will be easily perceived by others.

I asked several people (both women and men) what femininity means to them. Here is what they answered:

- that SOMETHING that a woman emanates.

- that sparkle that every woman has, naturalness, soft elegance, charm.

- the naturalness and personal tenderness of each of the characteristics attributed to the woman's identity, having as component elements such as gentleness, patience, kindness, sensitivity, openness to those around you, caring about the other, etc.

- those qualities that have to do with the mystique (that transcends charisma) that a woman can create around herself, with the creativity of all the feminine qualities that are awakened in a woman.

The conclusion I reached is that the more a woman has such awakened qualities and trusts in her, the more feminine the style, the elegance, the mystery, the more the surprise continues, something full of charm, something that sometimes takes your breath away, something magical and fascinating that has inspired many artists over time, e.g., *Mona Lisa* or *The Naked Maja*.

Every woman is unique and has her own specific beauty. But it is also true that some women are more aware of their endowments, strongly believe in them, and emanate with such power the beauty of their soul that their presence cannot be overlooked. Self-confidence gives a lot of charm to a woman. It makes her special. There are women who inspire us through refinement, sensitivity, compassion, devotion, firm faith in an ideal, courage and tenacity to fulfill it, through love for God and human values, through sensuality and naturalness, creativity, spiritual aspiration, through the power to overcome difficulties, through dignity, intelligence, sacrificial power or patience.

These women make us believe that it is possible to complete the manifestation of femininity or at least some of its aspects. They make us want to be better, more aware of ourselves. They make us believe in our talents and value them. It inspires us to cherish our lives with all its gifts and to live it to the fullest.

Unfortunately, not all famous women today are truly worthy of much admiration. In the case of some of them, the celebrity is much higher than their quality as women. We live in a time when in many situations advertising does not reflect reality. But there have been and still are great women worth finding out about. Some are less well known, although what they have achieved is admirable.

This leads me to the point I am making. Just as quantum science has put a stop to the myth of the genius in outer creativity since anyone can be creative using the creative process, it is time to bury the notion of the specialness of these "genius" women and replace it with the quantum notion: any woman can achieve this woman power by engaging quantum creativity in the inner dimension, that is, inner creativity. The arena is different. For men, the arena is intellect, that is their forte; emotions are important but secondary. Women's forte is emotion; so, women have to prioritize vital creativity, raise energy from lower (the first two chakras) to the higher chakras; mental creativity is secondary.

First, take the popular examples. Marilyn Monroe is a model of sensuality. Isadora Duncan impressed her audience by the passion with which she danced, by the way she transposed the music into motion, by the way she felt and expressed her soul through the music and the dance. Alexandra David Neel was noted for her unique writings on oriental spirituality, her books and works being absolutely fascinating. Aimee Mullins had her legs amputated from her knees, but this did not stop her from becoming a successful model, actress and sportswoman. At the awakened feminine level, there is no difference between Emily Dickinson and Eleanor Roosevelt, they both were expressing their soul.

Second, consider women at the soul and spiritual level. Mother Teresa impressed an entire world with her compassion and tenacity. *Amma Amritanandamayi Ma* is a special woman nowadays who impresses with love and self-sacrifice.

Somewhat before our time, St. Teresa of Avila is a model of devotion and spiritual aspiration. Less known, but still exceptional, is Lalla, an Indian woman who dedicated her life to communion with God. Sarada Devi (consort of Ramakrishna) showed an overwhelming maternal love for all human beings, which is why she was nicknamed "the holy mother of India".

Millennia ago, there is the case of the Tibetan mystic Yeshe Tsogyel (consort of Padmasambhava who brought Tantric Buddhism from India to Tibet). She was raped by seven bandits. How did she deal with it? With unimaginable resilience, she persuaded each of these bandits to become her disciple!

And still, when you take a look at history, it seems that the geniuses touted even for inner creativity are overwhelmingly men. Notice I had to introduce Sarada Devi and Yeshe Tsogyel via their male consorts. That is how people would recognize them. Great inventions and discoveries, fascinating works of art, magnificent spiritual/religious and political achievements are all marked by the contributions of male names. Is the genius a specific quality to the solar sex? Genius women are much less in number and when we hear about them, we ask ourselves if they really belong to this category.

Being a genius means you can't live happily unless you give birth to the work that you feel growing inside you every moment. No rest, no respite, the genius finds an endless energy for creativity. Countless hours they spend working and the patience which is part of his/ her being shows that the genius lives united with his/ her work: can't live without it. The genius finds in inspired moments of insight the simple and ingenious solutions to problems that concern all humanity and has all the skills necessary for the perfect production of his/ her work—outer or inner.

A genius marks the era in which he/she lives. Through his/her contribution, anything that occurs after such a human being can't remain as it was. They "disturb" the universe, to use the Physicist Freeman Dyson's word.

Any man or woman can achieve these performances, if they make all the appropriate efforts. Obviously, being a woman does not need to be an impediment. And yet, why are so few *recognized* female geniuses?

A handy observation is that **women give life**. This not at all a negligible aspect of a woman's life; it makes her less interested in creating a new work apart from her child. She gives life and devotes herself with all her energy and instills in that baby everything she thinks is the best, for which she often uses creative intuitions and insights. From this point of view, any real mother has already the vocation of a genius.

Sure, the qualities of that child have their importance as well. And if the child turns out to be also a real valuable human being, the fulfillment that the mother knows will have no resemblance. And the good her son/ daughter will make to the world will create a real difference, although her contribution as a mother remains anonymous.

But again, there are women who are geniuses and who create brilliant works. With their specific discretion and delicacy, their work marks the world. Why they are not always recognized must be attributed to male domination of human cultures until recently.

The choice is yours. Be a Mother—live your genius through your child; or fulfill women's mystique: awaken your soul. Or, do both. A good compromise is to do both: motherhood when you are young and explore your femininity at the mid-life transition, when as the fable goes, "the kids grow up, the dog dies, and maybe even the husband divorces you."

Femininity is a special gift meant to bring us happiness and fulfillment. It contains love, tenderness, feminine power, sensuality, inspiration, passion, creativity and other treasures worth discovering. It's such a beautiful gift that deserves to be celebrated every moment. In the whirlwind of everyday life, we are not always attentive to this treasure we women have received. We do not listen carefully enough to our feminine soul, its needs and aspirations. Sometimes we put our emotions and passions into a corner of our heart because the responsibilities and hardships of life grab all our attention. And we believe that by giving in to our emotions and sensitivity we cannot move forward in life without being hurt. But sensitivity and emotions are part of us, and it is necessary to accept and understand them. It is also natural to want to live the experience of love in its many forms of manifestation. It is natural to enjoy our sensuality and creativity.

Being a woman means more than being a mother, a wife, a caring daughter. It means more than dressing nicely, styling and makeup. It represents more than dedicating ourselves to our professional life, even if we achieve great recognized achievements in this field. Step boldly into the universe of your femininity. Discover here the multitude of facets of the Feminine. They are right in you and waiting to be discovered.

In a world of competitiveness, sometimes we inhibit our emotions, sensitivity and vulnerability. We tend to be extremely rational. We believe that in this way we protect ourselves, analyze situations "coolly" and make the right decisions. At times, rational decisions are useful and even necessary. But if this way of being takes over our whole life, we lose touch with the heart and its wisdom. To be a woman means to live in the heart much of the time. Give way to the wisdom that springs from there and defy the rational mind. Listen to your intuition, needs, dreams with the heart always active. Inspire those around you with your loving presence.

Where does Your Power as a Woman Come From?

It comes from self-knowledge, from discovering who and what you really are. It comes from self-confidence and understanding of your purpose as a woman. Your power as a woman comes from your power to love and give without neglecting yourself. It comes from your conscious connection to the creative force in your being. The woman is gifted to create life. And this does not just mean giving birth to a child, but filling everything she does with life, love and beauty. To inspire those around her with what she is.

And first of all, your power as a woman comes from the conscious connection to the essence of femininity which is the Archetype of the Eternal Feminine.

The first step on the journey would be to make a list of all your qualities as a woman. Write down your flaws as well. There is no point in denying them. If you recognize them and accept them, you can transform more easily. It's good to know where you're coming from.

The Force of Awakened Shakti

In the Tarot, it is the Blade no. 11, which evokes the Force, and symbolizes the colossal power of femininity. For one with awakened SHAKTI, this power becomes capable of taming and dominating animality (via the building of positive emotional brain circuits balancing the negative and developing femininity as a character trait).

*Figure 4 Durga: The divine mother - the archetype
of feminine spirituality*

The mysterious force of the fascinating woman who has fully awakened the state of SHAKTI (fig. 4) is unlimited because in this way she connects herself to an endless sphere of force that exists in the macrocosm. The lion that has strong paws is right near her. The strong and powerful mouth of the lion is now opened with the help of the gentle hands of the woman who has awakened the state of Shakti. The lion has strong fangs, but it will yet be unable to bite the woman who has awakened the state of shakti , because she keeps its mouth open and the lion cannot resist her. The mysterious force of Shakti that she manifests is of such a nature that it is an easy game to subjugate the mighty lion and to ceaselessly keep it under her irresistible influence.

This force of Shakti is undoubtedly the strongest force in the world. You have to understand something important: *this force of Shakti is not a physical Newtonian force that causes a body to change its state of movement; in quantum physics, material forces are good for producing possibilities for consciousness to choose from. In the mysterious force of Shakti, is revealed the power to persuade someone's choice using the energies of love.*

Love is the endless mysterious energy that comes from God and leads us towards Unity. When the persuasive power of measureless love manifests itself in the human, she (or he) gets to love the entire universe. For this reason, the woman in whom the state of Shakti awakens feels one with everything that exists and has a new life. To Her, love is life, love is the food of her being, love is God, the eternal.

Questions to ponder upon

What is in reality the role of mysterious feminine, which most often is not understood?

What was in reality the role of the mysterious feminine in the case of the many renegade followers of some spiritual paths, who ran away but came back to spirituality and soul-making?

What was in reality the role of the mysterious feminine that the greatest wise men and clairvoyant of the Indian Vedas were comparing with the sunrise (the *Gayathri mantra*), with the immensity of starry sky, with the paradisiacal shinning, with the celestial beauties and with the state of ideal purity, that no gods or people know in its essence?

To talk about the *inner* feminine is almost a non sequitur, because through its biological definition the feminine is something inside and secret. To dive within the soul and within our inner universe—does that not mean we will be meeting our feminine? Is it not for this reason that the majority of so-called macho men is stubbornly cramped by remaining to the surface of their being, accommodating an embracing state of superficiality, afraid not to lose somehow their cultivated masculine identity?

I will give below three exemplary Women, two from the West who were crucial in bridging the East-West divide, a divide created by the rational mind of Western men. The third example is from Tibet. I find such stories to be inspiring for the feminine soul. Here is a compilation of what I found about them and their life in various sources.

"The Mother"—Aurobindo's Disciple, the Mother of All

Mirra Alfassa, later dubbed the Mother, was born in Paris in 1878. Her father was Turkish and her mother Egyptian—both Jewish. Since the age of 5, she realized that she does not belong to this world and started a spiritual practice following her own inimitable way, as she saw fit.

She began to enter the state of bliss while sitting simply at the table, to the dismay of her family who could not understand her. At the age of 12 she was already having out of the body experiences. At age 16 she enrolled in the School of Fine Arts, where she received the name "The Sphinx", which she subsequently used to sign her exhibitions. Between the age of 19 and 20 she achieved a conscious union with God (state of samadhi) without the help of any master or any book. Shortly after, she discovered Vivekananda's book on *Raja Yoga* of meditation and samadhi which helped her to make even greater progress.

These spiritual achievements did not impede on her normal social life. At the age of 19 she got married and had a son. She became a part of the circle of the contemporary artists; she was a good friend of luminaries such as Rodin and Manet.

In that period, she discovered the Bhagavad Gita and fell hopelessly in love with Krishna. When in her dreams the image of an Asian male figure began to appear, she thought she was seeing Krishna, but in a few years, it proved to be her spiritual guide, Sri Aurobindo.

In 1908, Alfassa divorced and focused all her energy toward spiritual quests. Together with her brother Matteo, she established a spiritual group called The New Idea, in which she met every Tuesday evening with people interested in psychic phenomena and mysticism. In parallel, she founded a group called Thinking Women United, where Alexandra David-Neel was also a member and with whom she became close friend.

In 1910 she had an experience of awakening, where she realized that the Divine Will was at the center of her being and ever since that moment she was no longer interested in any personal desire, but only to follow that Divine Will.

In 1912, she organized another group of about 20 people, called Cosmic, whose purpose was the attainment of self-knowledge and self-control.

During this period, she met her second husband, Paul Richard. He traveled to India and had met Sri Aurobindo, with whom he was corresponding. In 1914, Alfassa and Paul made their first trip together to India, where they met the master. Alfassa recognized Aurobindo as the one who appeared to her in visions. At that moment, her mind completely shut down to the external world for a few days and she remained in an expanded state of cosmic consciousness.

During World War I she fled to Japan, where she met the poet Rabindranath Tagore. A Japanese friend wrote about her: "She came to learn the language and integrate with us. But we had so much to learn from her, she is so charming and unpredictable." Finally, she returned to Pondicherry, where Aurobindo taught, and she settled there in 1920. Once the master retreated into relative isolation, in November 1926, she became the head of Aurobindo ashram. Sri Aurobindo believed that the Mother was an avatar of the Supreme Shakti, and he honored her as such.

It is said that Mira was encouraged by Aurobindo to wear saris, the reason she had a collection of about 500. When someone was offering her 100,000 rupees for one of the sarees, she sold them all, with all the accompanying ornaments, thus gathering funds for the ashram which was having financial difficulties after the death of their spiritual guide. She continued the work of Aurobindo by founding Auroville, a city where "people can live away from national rivalries, social conventions, perverted morals and hostile religions".

The 1,500 people who live in Auroville today live in houses that they have built themselves, cook with solar energy and trying to give a new meaning to the concept of community. Their settlement is a laboratory, and they are the pioneers of human evolution. In Auroville, there is no system of laws, police or mayor. Life is entirely organic and television and similar entertainments are completely missing. Many believe that this is the place from where there will be a start of the regeneration of a new humanity.

"The Mother" as her name remained, died in 1973, leaving behind an impressive bunch of work, including dozens of books with the testimony of her experiences on various spiritual topics. Was Mirra Alfassa a soulful genius? No one can argue otherwise. She had the endorsement of a high spiritual being, and her work is the perennial testimony of her genius, while it is continuing to transform the world.

Lilian Silburn, the Mysterious Knower of Secret Teachings

Kashmir Shaivism would not become so easily available for us all today without the translations of the incomparable Lilian Silburn, a French woman whose presence in the world remains one of an almost frustrating discretion. Little it is known about her personal life, but her work shows that she was a genius. She translated and commented on the hymns of Abhinavagupta (the father of Kashmir Shaivism), translated his book *Vijnana Bhairava Tantra* and wrote a reference book about Kundalini energy. Her doctoral thesis, presented in 1948, is called *Instant and cause. The discontinuity of philosophical thinking in India*; and is a journey full of common sense into the Vedas, Upanishads and Buddhism. In all her writings she proves that not only she is an erudite thinker—but also that she explored consciousness via her own experiences.

It seems that her spiritual guide was a certain Radhamohan Adhauliya, who was not a Shaivite but a Sufi. She was Hindu and yet she received the teachings from this spiritual guide who was known to be very orthodox in his thinking. The way she integrated the two traditions, to which it was added the Shaivite influence, continues today through the spiritual aspirants whom she influenced until 1993, when she left this world.

She led for many years the National Center for Scientific Research in Paris. Over time, she became known in the West as the greatest authority on Kashmir Shaivism, a field in which she published numerous studies and articles.

Her most important book, *Kundalini—energy from the depths*, can be downloaded for free from the internet. There are no references in this book about the author; instead, there are many passages about her experiences. It is a book that can transform your life. If we take into consideration that Lilian Silburn is the one that has exposed for the West one of the most sophisticated spiritual wisdom traditions of the planet— Kashmir Shaivism—we can be sure that she was a truly remarkable woman.

Yeshe Tsogyel's Inspiring Story

Yeshe Tsogyel's story is partly mythologized but the historical part is easy to discern from the following account which include both history and mythology.

Yeshe Tsogyel is a remarkable and inspirational female figure and her life is a source of confidence and aspiration for those who want to have an authentic and at the same time tangible purpose in their existence. She was a perfect tantric master in her own right as well as one of the main mystical consorts of the great Padmasambhava.

There are 4 great Tibetan Buddhist traditions, with many yogic disciples (men and women alike) and masters of these paths. All the teachings of Tibetan Buddhism appear in the legends of the life of Yeshe Tsogyel. Buddha taught in India and prophesied that the future would see the birth of a great teacher in Danakosha, a territory now in Afghanistan.

That great master was Padmashambhava. In his prophecy, the Buddha stated that this teacher would have a special influence and would revive the Tantric tradition. Before the tantric teachings of Padmashambhava, this spiritual path was practiced by few followers. It was not until his work that it became widespread. It is said that Buddha Shakyamuni taught mainly via the intellect and practices based on his Sutras, and Padmashambhava taught via experiences of vital energies, the subject of the tantras. Since then, both types of spiritual teachings have spread widely throughout the world.

According to Buddhism, there are 4 great ways in which beings can be born into this world: from an egg, from a woman's womb, from a heat source or energy source, or in a miraculous way. The first two are physical, the last two metaphorical.

Padmashambhava became a seer when his kundalini rose from the vital energy coiled up in the low chakras and opened his chakras like the opening of lotus flowers. Hence it is said he was miraculously born from a heat source in the middle of a lotus flower.

At that time there was a king named Indrabodhi, who had no sons. He had made many donations to the poor in order to earn the right to have a son. Indrabodhi was blind, and the prophecies circulating in his day claimed that with the coming of his son, he would gain his sight. When the king heard that Guru Padmashambhava was born in the middle of a lotus, he rejoiced and went there with his whole retinue to invite Padmashambhava to his palace, proposing that he adopt him as his son. It is said that making this gesture, the king began to see. Padmashambhava succeeded Indrabodhi to the throne and served the people for 108 years spreading Tantric Buddhism.

Yeshe Tsogyel was also born to support the spread of Tantric Buddhism in Tibet. Many auspicious signs accompanied her birth. Then, once born, she was fed from a breast-shaped sandalwood trunk. Yeshe grew up much faster than other children, developing in a month as much as others did in a year. When she played with children her age, Yeshe left the imprint of her palms and soles on the rocks. She could read without anyone teaching her. When her father asked her how she was doing this, she replied, "I am Yangchenma" referring to her previous incarnation.

Already a boddhisattva, it is said that Yeshe could be anywhere and in any form she wished to take. She had reached the supramental state of consciousness (mature soul). Yeshe showed great compassion, helping the poor. Her mind was sharp, and it allowed her to understand everything she set out to do. Yet, her character was peaceful, calm. Her spirit was very compassionate, and everyone around her became calm and happy.

Yeshe used to meditate a lot. Through meditation, she perceived the true nature of existence. At the age of 7 or 8, Yeshe already prayed for the well-being and happiness of all beings. She wanted to achieve that knowledge and wisdom that would allow her to help others as much as possible. At the age of 13, many nobles in the kingdom began to come to ask her for a wife. But her parents, aware of their daughter's extraordinary destiny, did not comply with the request of any suitor. All Yeshe wanted was to attain the wisdom of a Buddha. Her popularity, the signs presented at birth, and her great compassion gradually began to be known everywhere in Tibet. Eventually, they reached king Trisong Detsen's ears. He sent one of his ministers to Yeshe Tsogyel's parents, asking them to bring the girl to court.

Hearing this, Yeshe fled the house, and took refuge in an abandoned place, where she took out all her jewels and threw them in the dust of the road, scattering them in ten directions. She pleaded with Buddha and bodhisattvas to remove any obstacle to her enlightenment.

While she was praying like this, a 16-year-old man suddenly appeared next to her, telling her that it was useless to cry or throw jewelry. On the contrary, in order to achieve the goal she desired, it was necessary to have an unwavering will, and to continually implore the bodhisattvas and Buddhas. "Your prayers will be heard in this way, and your wishes will be fulfilled," the boy added. "Come with me, and I will show you the way to enlightenment." Then he took Yeshe's hand and they instantly traveled to a remote, deserted green and beautiful place. He advised her to stay there.

The boy, a manifestation of Guru Padmashambhava, taught Yeshe Tsogyel certain teachings about life and *Samsara* (the illusory world). Yeshe asked him how to practice after he left; the young man taught her how to practice, as well as all that was necessary for her to know about the nature of consciousness.

Then she asked him who he was and where he had come from. The boy replied that he came from *Dharmakaya* (consciousness in suchness) and insisted that she practice everything she had learned from him. Yeshe begged him to allow her to stay with him forever, to continue to receive his teachings. But the young man replied: "The time will come when we can be together. But now I can't stay long, because I'm just an apparition." Then he disappeared.

Yeshe Tsogyel felt sad and happy at the same time. She wondered if it had all been a dream, or if it had happened at all. Then she thought it couldn't be a dream because she was awake. The enchanting place she was in brought her much happiness in her soul.

But, Yeshe thought, there was no food or clothing there. Then she saw some wild animals, and said to herself, "If they can live here, I can." So, she survived in that place, feeding on plants and quenching her thirst with brook water. If it rained, Yeshe took refuge in a cave, where she could meditate. If the weather was fine, she lived in the middle of nature.

At the same time, Yeshe continued to practice and, as time went on, she reached the threshold of wisdom. All this time, her parents were looking for her, worried about her fate. They held the visiting minister responsible for the girl's disappearance, asking him to get involved in a search for her.

Later, the minister returned to his king, telling him everything he had seen and heard. The king sent many of his subjects throughout the kingdom to seek out Yeshe, promising them great rewards. Yeshe continued to evade them and meditate in her retreat.

One day, she was spotted by pilgrims passing through the area. They were amazed to see such a young and beautiful girl meditating in a secluded place, so they asked her where she was from and what she was doing there. Yeshe thought that if she told them the truth, those pilgrims would take her back to her family.

So, she told them she didn't remember who she was and where she came from. But the pilgrims did not believe her. They even accused her of being a nun who had run away from a monastery, to escape from the harsh rules there, and to practice as she pleased. The pilgrims also asked Yeshe what she was eating, and were amazed to learn that wild plants were enough food for her.

Eventually, she taught the pilgrims how worldly desires, including those of a chosen food, attach the self, sinking it into Samsara, the world of illusions and suffering. She spoke to them about how uncontrolled emotions sustain ignorance, and in turn are born of ignorance.

The pilgrims, deeply astonished, offered her gifts, so that they too might accumulate some merits through this offering. Then they left Yeshe and went on their way. But they kept talking everywhere they went about this unusual girl.

The minister referred to above heard this news, and he also went to the place where he had heard that the special girl lived. When he found her, he offered Yeshe to come with him to the king, insisting that the king was not an ordinary man, but behaved and looked like a god. The minister then proposed to Yeshe that she be the queen of Trisong Detsen. But the girl refused to put any price on his offer, and asked him to let her meditate in peace, wherever she was. It was obvious to her that a queen's life would bring her nothing but the accumulation of sins and suffering. The practice of dharma in that lonely place seemed to her a much more attractive prospect!

The minister became angry and continued to remind her that she had arrived there without her parents' approval. Then he threatened her that if she did not come willingly, he would find other ways to force her to follow him. Seeing that the girl remained unyielding, the minister then forcibly took her from her place of meditation, dragging her to the king.

Yeshe was only 13 years old when she became the queen. King Trisong Detsen invited Shantarakshita and Guru Padmashambhava, as well as other great sages of India, to come to Tibet. The king received the tantric teachings of Guru Padmashambhava, and as a sign of deep gratitude, Trisong Detsen offered all his possessions and even the queen herself to Padmashambhava.

The Tibetan tradition speaks of a certain form of divination, in which a disciple throws a flower on a mandala; if that flower falls in the right place, the disciple can receive the teaching related to that mandala. Yeshe Tsogyel thus received the Vajrakilaya teaching. Padmashambhava told her to practice it, gave her the secret spiritual name of Dechen Gyalmo, and advised her to practice meditation in a secluded place. Then he predicted that she would obtain that extraordinary *siddhi* (paranormal power) of being everywhere in any form at will.

Although she had already acquired that achievement in her past life, Yeshe continued to practice being in thousands of other places of retreat to bless and permeate them with holy resonance. Thus, future practitioners who meditate there will not face major obstacles in their path.

Yeshe also obtained the siddhi to remember all the oral teachings of Guru Padmashambhava. Also, through diligent practice, she obtained the diamond body, which remains unchanged over time.

After attaining enlightenment, Yeshe devoted all her actions to the purpose of saving other beings. The first thing she did in the service of others was to free the evil minister from negative karma, who had left the physical world and going through the after death bardo of hell.

Yeshe projected herself out of her body into that infernal world and released not only that former minister, but other beings there as well.

Then Guru Padmashambhava sent her to Nepal to save an acharya, a scholar. It was a journey that was particularly difficult at the time. On her way, Yeshe met 7 robbers, who, upon seeing her, planned to attack her and steal her belongings. As they approached her, Yeshe saw them as deities, and saw herself offering them jewelry. Thus, maintaining this attitude, she offered them what little she had.

When the robbers heard her pleasant voice and saw her beautiful appearance, they thought of raping her. In her great compassion, Yeshe allowed them to have sex with her. As a result, those robbers were freed from all their obscuring behaviors and negative karma. When they realized Yeshe's compassion, they deeply regretted their deeds, and begged her to forgive them and became her disciples. Then those robbers really set out on the path of virtue.

While Yeshe was wandering through Nepal, she met a young man who asked her if she was looking for him. Yeshe agreed he was the scholar she was looking for. So, she accompanied the boy to his parents, and asked them to take her in their home. However, the boy's parents only allowed her to stay in a tent near the house. Later, when those people heard her sublime voice and perceived her benevolent nature, their souls were filled with devotion to Yeshe, and they asked her if she was a human or divine being. They added that it was dangerous to move around alone. The boy's parents also noticed that their son had some kind of karmic connection to Yeshe, and begged her to stay in their home as their son's wife. She did not resist but demanded that the boy was the one who had to follow her to Tibet to receive the teachings of the great master Padmashambhava. The boy's parents did not accept this condition.

Yeshe insisted, saying, "The boy should come with me to Tibet. This is my master's command, and I obey them. You can keep all my belongings. Guru Rinpoche never told me to stay here, and therefore I will not stay. If your son comes with me, he will achieve the state of enlightenment, and you, in turn, will be free from suffering."

The boy's parents said to her, "If you can give us as much gold as our son weighs, then you can take him with you to Tibet." They thought this was an impossible task for Yeshe. But she agreed and then went in search of gold. On her way, she met a family whose son had just died, and they were taking his body to be burned. Yeshe felt a lot of compassion for those grieving parents, and she thought about how she could help them. She went to them and said, "If I can restore your son's life, what can you give me in return?" The father of the deceased boy replied to Yeshe that he was ready to give anything for his boy's life.

Yeshe asked for the weight of that young man's body in gold, and the father agreed. Then Yeshe meditated for a long time and prayed to Padmashambhava, and at last the body of the young man who had left this world came back to life: beads of sweat appeared on his body, the color returned to his cheeks, and at last the young man opened his eyes, calling out for his parents.

Overwhelmed with happiness, the parents complied with Yeshe's request, giving her the gold she needed. Meanwhile, news of her ability to restore the young man's life had spread throughout Nepal. When she returned the promised gold to the young acharya's parents, they fell at her feet and told her they did not want the gold. "It was not greed for riches that made us ask for this gold, but too much love for our son, whom we did not want alienated. So, keep the gold and take our son with you!" They said.

Yeshe said that birth as a human being is far more precious than gold, and so she had no reason to keep that gold either. Then she bade the young man's parents goodbye and took him with her. Together they went back to Guru Padmashambhava, who offered them all the tantric teachings that, by practicing together, they were able to attain to the state of enlightenment the Tantric way.

Yeshe is said to have lived to be 200 years old. It is said that at the end of this period, she turned into light, and passed directly into the Pure Land of her Guru-Rinpoche Padmasambhava. You can see that all this about Yeshe's life is part history part myth, but much of the meaning of the myths can be accommodated using quantum science. We will end Yeshe's story, however, with a purely mythical tale, just for inspiration.

The Padmasambhava couple's fame grew so much that Brahma—the creator of the universe—himself is said to have been amazed. He heard that a second Buddha had been born, and he and his consort were teaching the spiritual teachings in a new way. He wanted to see if this couple was spreading the teachings out of compassion, or with some personal gain in mind.

Thus, Brahma decided to test the mind of Yeshhe Tsogyel. He descended to earth, disguised as a leper, and came to the cave where she meditated.

Yeshe heard the leper's cries, and her heart leaped with compassion. She told him: "I look at your wounds, and I realize how much it hurts. But the shouts don't help you; more important is to look for the cause of the pain, which lies in your previous, negative actions. The three poisons are the cause of your suffering, and you must avoid them. You must confess your past mistakes, and I will look for a way for you to overcome these sufferings."

Brahma replied, "I met many people who promised to help me, and they were unable to offer any help to me. So, I don't think you'll be able to help me either. Better leave me alone!" Yeshe went on to say, "I want to help you, not to hurt you, so why should I leave you?"

Brahma also continued: "I want you to leave, because my ex-wife looked exactly like you; she passed into the afterlife a year ago. Your image makes me suffer both physically and mentally. My pain can only be overcome if I am offered a kneecap. But no one will sacrifice a part of his body, so there is no point in staying around me."

Yeshe felt endless compassion for that leper. She told him, "You should realize that Samsara is suffering so you will want get rid of attachment to it. If you realize that your suffering stems from your previous actions, and you repent, all your negative karma will be nullified. But if giving a ball for your knee can really help, I'll give you my own ball."

The leper said, "Offering to give me your kneecap makes me really happy!" Then, he cried, "Your kneecap can make me overcome the physical pain, but how can I overcome the pain of losing my wife?"

Yeshe replied, "Once a being passes into the afterlife, it can no longer be physically accessed; therefore, it is better to forget your wife. "

The leper replied, "Since you look so much like her, if you agree to be my wife, it would take away my pain. But since you wouldn't do that, you better leave now and leave me as I am. "

Yeshe, in her boundless compassion, agreed to be his wife.

The leper was very happy. He first asked her to give him the patella. Taking off her kneecap, Yeshe fainted. But when she recovered and opened her eyes, she saw Brahma in his divine form before her. She asked him where the leper had disappeared.

Then Brahma, very pleased with her attitude, apologized to her for the test she was given, telling her that he was ready to do whatever she asked him to do to help other beings. Yeshe then asked Brahma to reveal how to remove obstacles and achieve enlightenment in a way that everyone can do; in this way enlightenment would be accessible to all beings.

The conclusion of the story: that is how the creative process became known to humanity, which was codified centuries later.

The Archetypal Basis of Femininity

Celebrating our Femininity and aspiring to be a Woman with awakened transformative power, means looking at our femininity as a special gift that Life has given us, and put it in practice in service for the world. It means discovering and manifesting the treasures of our feminine soul.

It means to (re)discover our sensitivity and feminine power, creativity and the power to love, the courage to be ourselves and to affirm our uniqueness. In the soul of every woman there are all the seeds of Femininity. Some have already sprouted and borne fruit, bringing fulfillment and satisfaction for a fulfilled purpose. Others are waiting patiently to be discovered and brought to light. They are waiting for the moment when they can bloom and spread their perfume. We call these seeds of femininity Feminine Archetypes.

It is always good to know where you want to go. The next step is to get some landmarks. Femininity is very complex, so it's hard to grasp in its entirety right at the beginning. Therefore, it is useful to divide the journey into different shades and stages of it. Invoke the different female archetypes as appropriate for a particular stage. They can be your guiding light at each stage of your path to fulfill femininity.

The Feminine archetypes are the different aspects of femininity in suchness. Here are the main ones in Jungian representations: The Mother, The Ideal Lover, The Heroine, The Sage.

Take this idea with you: Every Modern Woman can blossom into a wonderful State of Awakened Shakti.

A soulful woman (aiming to be wonderful at all levels) naturally becomes a perfect mother, wife, sister, daughter, friend, and professional. Her awakened self-confidence, intuition, mental power, strengthened by the power of her great love make miracles in the education of her children, in her social and professional relationships, in her intimate bonding as a couple!

Because of all the qualities (only partially delineated above) such a woman of Shakti will manifest amazing efficiency and creativity in her work (thus be great in social integration); she will experience very happy relationships (intimate and familial); her spirituality will offer answers to all her questions, and her destiny will start to unfold its mysteries towards the accomplishment of her dharma (the very foundation, meaning, and purpose of her life).

Self Evaluation Questionnaire (for Women)

I will end this brief introduction to women's spirituality with a questionnaire. Use it to test where you are in your journey. And if the quantum scientific approach to what is potential for today's women has impressed you, engage, explore, and embody Shakti in your body's selves.

1. Are you happy to be a woman? Why?

2. What importance has your physical appearance to you?

3. On a scale of beauty from 1 (low) to 10 (high), where would you place yourself?

4. What are your three biggest qualities?

5. What are your three biggest faults?

6. Are you involved in a couple relationship?

7. On a scale of happiness from 1 to 10, where would you place yourself as a couple?

8. On a scale from 1 to 10 how important is it for you the erotic fulfilment?

9. What does 'woman-femininity' mean to you a) before reading this chapter and b) after?

10. Describe the "the ideal man" for you.

11. How would you like to become as a woman?

12. What are your next goals in improving your life from a feminine point of view?

EPILOGUE

The Re-enchantment of the World at Large

How do we re-enchant the world at large to the way some of us experienced it in our childhood? Can we? We must. The dynamic duo of ignorance —religion and materialist science– both fed by pure rationalism have exacerbated our usual social challenges to the current crisis condition. The only solution, albeit long term, is integration of the two worldviews and subsequent dissemination. The first is seeing the end of a promising beginning. It is time to end this book with some discussion of the second part of the solution.

In 1999, I (Amit) went to Dharamsala, India as part of a group of post-materialist scientists and other avant-garde thinkers to have a conference with the Dalai Lama. This is the conference I will always remember with chagrin because my friend Fred Alan Wolf and I got into a fight as to whose fine prints of idealist interpretation of quantum physics is the right one. Very soon, all of the participants were taking sides. When the organizers complained to the Dalai Lama though, he laughed and laughed, diffused the fight, and everybody including Fred and I were having ecstasy.

The Dalai Lama did not end there. He turned serious and asked us to use our energy applying our idealism to solving societal problems. I was touched by Dalai Lama's sincerity and decided to do the same. A turning point.

My first application of quantum science was to health and healing, to develop a theory of medicine integrating the purely materialist approach of allopathy and the many old/new approaches of more holistic healing that go under the label of alternative or complementary medicine.

You know, allopathic medicine is largely an evidence-based medicine; what this means is that the pharmaceutical medicines that you take for curing disease are by and large the result of clinical studies; there are no compelling theory that guides the research. In this way, allopathy is really halfway science with just one prong rather than the customary two prong approach that is the power of science.

In contrast, alternative medicine systems have some theory behind them except that they theorize non-material bodies and non-material energies of healing that allopaths look upon with suspicion.

I began with the conviction that since quantum science has found the science behind non-physical bodies of our software, an integration of the conventional allopathy and alternative medicine should be at hand. I did some research and published the book *The Quantum Doctor* about my efforts to integrate.

It took me many years more and the collaboration of a physician my co-author Dr. Valentina Onisor, before we finished a reasonably good stab at an integrative theory of medicine. Read our upcoming book, *Quantum Integrative Medicine*.

In 2007, I was in Brazil before an audience speaking about the quantum integration of science and spirituality. A young man challenged me, "Where is the scientific evidence for God? Does it even exist?"

To this I said intuitively, "Young man, scientific evidence for the existence of God is already here, the question is, what are you going to do about it?"

I don't know the effect of that outburst on my questioner, but I myself was startled hearing that pronouncement. What am I doing about this great information which can re-enchant the world, reconnect humanity? Thus was born the movement called quantum activism: as you try to change the world as activists try to do, you must also change yourself with the help of the quantum principles.

Soon, two amateur film-makers Ri Stuart and Renee Slade made a documentary of my idea of quantum activism called *The Quantum Activist*. Earlier I was featured in a movie called *What the Beep Do We Know*? Coming back to back after that film, the documentary was a success. The movement got some traction, especially in the USA and Brazil. I began to give 5-day workshops on quantum activism with the help of my friend the enneagram teacher Adriano Fromer.

What I found is that a five-day workshop is good enough to inspire people, but it was just a workshop high. People go workshop-hopping as a new hobby to seek inspiration and expansion of consciousness, but it seldom goes deeper. In the few cases though it does go deep, and the numbers were adding up and I began to see the need for a more committed higher education for these people.

Today, what people get as higher education is "higher" only in the sense of preparing you for a higher-paying job, nothing more. How far we have fallen from an enchanted past when higher meant higher happiness and higher intelligence, when education itself was geared to remove ignorance about the enchanted reality of the soul and the spirit!

I was determined to provide real transformative education with soul-making the goal. Today, many politicians in the USA talk about America having lost its soul, but no one takes them seriously. The same is true all over the world. It is not a political point to score, though it is true. We must take it seriously, those of us who can. I have.

The task looked enormous at first: financing, building organization, finding a group of dedicated helpers, those were the first thing. However, what was really daunting is the problem of getting affiliation for such a degree program even at the Masters/PhD level when the field itself—consciousness-based science—remains controversial no thanks to opposition from materialists.

That Hollywood moving words "When the field is ready, people will come," kept ringing in my consciousness and that is basically what happened between the years 2016-2018. In 2018, we ran a pilot program with mixed success. Initially, we were promising only graduate level certification; the program was mostly online with ten days of proximity—workshop style. It is important for transformative training to be given in proximity; that of course, adds to the cost. Nevertheless, students came; the demand was there, and it was international.

The main problem—twofold—was still unsolved. Affiliation and finance. When some visionaries of a private fully government affiliated university in Jaipur, India, approved of our program, I did not hesitate. I decided to become a quantum businessperson and finance the operation myself.

*Figure 5 Dakshinamurthy: The archetype
of wise teacher situated in wholeness.*

Thus began QUANTUM ACTIVISM VISHWALAYAM. Vishwalayam means home of the world as our institution was intended to be. Our teaching model is *gurukulam*; learning under teachers who live their realized wisdom. The ideal of such a wisdom teacher is called *Dakshinamurthi* in Sanskrit (fig.5). I myself was humbled to accept such an ideal for myself; I knew I am not fully "there" yet, but I was ready to start learning by teaching. The same can be said about Valentina. My co-teacher.

Others—Anil Sinha, Adriano Fromer, Dinesh Kukreja, Terry way, Krishanu Goswami, a very able group of helpers—were there to support. We started our masters/PhD program in August, 2019. Our first proximity sessions were held in Jaipur, India in Jan 2020 with roughly about 40 students. It was an unmixed success. Our sponsor, the University of Technology, Jaipur, fully government of India affiliated, was pleased. The program is being taught with the banner of the DEPARTMENT OF QUANTUM SCIENCE OF HEALTH, PROPOSPERITY, AND HAPPINESS under UOT.

Learning by teaching is easy to conceive, not so easy to deliver. The first difficulty I faced was the inadequacy of the books I had already written and published (about ten books) for the task; students needed at leasrt twice as many, supplemented by as many video courses. How would I do that in a mere two years, the time it would take for our first Master's batch to graduate? So far, I have been a taking it easy as a writer, a book every two years.

What has been amazing in the last two years is an aborning capacity I have never known but somehow anticipated. When your intention is in consonance with the purposive movement of consciousness, things fall in place: intuitions come, vitality is awakened in previously inactive centers, helpers become available; I felt like I was just making the causal connection to make the books and video tapes happen.

In this way, we have now begun graduating master's students as promised. Our students are receiving transformative education that prepares them to live life in an enchanted level in addition to integrating their thinking, living, and livelihood. As they earn their livelihood using quantum science in any of the three professions of physical health, mental health, and business, they would propagate the new paradigm and with it a new worldview.

I have heard: one person can only expect to deeply influence about eight other people during a lifetime. Initially, the numbers are small, but they grow.

For the past four hundred years we have been educating people with material stuff of rationalism that has helped bring material prosperity but a dire poverty in people's internal life. In the name of secularism, we have banished value education and archetypes from our societies and cultures all over the world. This is not the place to argue about secularism, but it is important to recognize that spirituality is not religion, it is innate in humans, and it does not profess dogma. And now we have quantum science which theoretically and experimentally supports the spiritual foundation of human life. And more. It gives us a field-tested methodology—a technology of the spirit so to speak—in the form of the fully scientific creative process.

Can we re-enchant the reality we live? Yes, we can, and we must. The steps are clear; the science is there, the path of creative transformation is laid out in an unambiguous way, verified by empirical data, and the education will only get better as our graduates do their research and get their degrees and use the quantum science in their professions. For more details, read the appendix.

Sixty years or so ago, the great leader Martin Luther King shared his dream—eradicating racism in America—with us. We are still struggling with fulfilling his dream. When the World War II ended, many people said, Never again. We still struggle keeping peace in the world in a large scale. In the meantime, the situation has only gotten worse; and the worldview view battle between two faulty worldviews is the root cause. We hope this book helps you see that.

Let me share with you now my dream. Now we know we have an integrative quantum worldview that promises to bridge all our dichotomies. Now we know there is no free lunch. Now we know that we need to transform in order to veer the world toward Justice, unity, and peace. My dream is knowing all this, we will not fail in transforming the world. Hard work you say? When have we been afraid of hard work?

APPENDIX

Books for Further Reading

What does the transformative quantum higher education consist of that it can promise so much? Let's count the ways:

- Your education starts with a thorough examination of the questions addressed in this book: questions of reality, matter, consciousness, God, and soul.

- You learn the basic concepts of quantum physics and quantum measurement, quantum worldview as opposed to the current Newtonian one, and quantum activism. Additionally, you learn how consciousness enters science in an unambiguous manner, and the quantum principles of being, making changes, and relationships.

   Books to read if you want to continue self-help learning: Goswami. *The Everything Answer Book; The Self-Aware Universe.*

- Next comes the quantum science of manifestation—the know-how of how to create your own reality, enchanted if you so intend. Gone are the days of ambiguity, secrets, and undelivered promises. There is no free lunch; but you get what you are ready to strive for.

   Book to read: Goswami, Blake, and Stuart: *Quantum Activation: Turning Obstacles into Opportunities.*

- You learn the quantum science of reincarnation. Archetypal exploration takes much time, but reassuringly time is not the essence since we reincarnate again and again. With the prerogative to choose our own Archetype-dharma—we have the option of making each incarnation a bliss to follow. Learning about the nuances of reincarnation helps you to find your dharma.

Book to read: Goswami, *Physics of the Soul.*

- Curious about the lineage of the new science? Quantum science is a science of consciousness that begun 7000 years ago and kept alive in a small scale through ups and downs of civilization and finally finding fruition as a full-pledged science consisting of an unprecedented four prongs: theory, experiment, experience, and technology of transformed enchanted living whereas conventionally, science has three prongs max: theory, experiment, and material technology.

  Books to read: Goswami, *The Visionary Window; God is not Dead.*

- Developing a science of human experience has been the biggest challenge of modern science so much so that philosophers concede that this is the *hard* question facing science today. Quantum science has solved this gigantic problem. You learn the details here.

  Book to read: Goswami. *See the World as a Five Layered Cake.*

- Our experiences are mostly mediated by our brain. The brain, conventionalists argue is a material machine designed to be Newtonian. Is there a quantum brain? How does the brain manifest its ability to refer to itself as the self/subject of an experience? How does the brain acquire its preponderate negativity and pleasure/addiction? Our teachings help you to understand and answer these questions and show you the way out: developing the quantum brain which enables us all to transform brain's behavior.

- Yes, you can change and optimize your brain by learning to explore the quantum brain. This program tells you how. You also learn how to deal with brain's occasional pathologies, especially those coming from early upbringings.

Book to read: Goswami and Onisor. *The Quantum Brain*.

• The transformative methodology, the science of manifestation, all crucially depends on engaging creativity and the creative process. This is one of the greatest discoveries of the twentieth century science not recognized in materialist science only because of myopia. Quantum science explains all the unbelievable aspects of creativity that have confused people through the ages. Now you can learn everything needed for your creative exploration in a concise course. We have come a long way, dear reader.

Book to read: Goswami. *Quantum Creativity*.

• Materialists are long and vocal on the merits of Darwinism but keep quiet about its shortcomings of which there are many, the most famous being the fossil gaps. But then they misunderstand life itself and therefore the proper way to interpret Darwin's work. You will learn quantum biology of life and its hardware-evolution and software-development. You learn how to distinguish life and nonlife, learn how to put the concept of survival of the fittest in proper perspective, how to include purpose in life's evolution and development, in the way of learning a theory of quantum evolution that agrees fully with empirical fossil data. Finally, you learn how to join the adventure of the ascent of humanity to new heights of mental and vital development that includes the fruits of archetypal explorations.

• You will also learn how to use Rupert Sheldrake's great idea of morphogenetic fields integrated within a science of feelings and emotions and the chakras—centers of feelings in the body developed in India millennia ago.

Book to read: Goswami. *Quantum Biology and the Ascent of Humanity*.

- The quantum science of experience validates the millennia old idea that we have many bodies; the material hardware must be supplemented by nonmaterial software accounting for these extra bodies. Therefore, our healthcare must involve both the care and healing of all our bodies—hardware and software. You will learn the integration of the conventional and alternative medicine, a great synthesis of modern times that is taking place.

  Books to read: Goswami. *The Quantum Doctor*; Goswami and Onisor. *Quantum Integrative Medicine*.

- In the same vein, you will learn how the different current models of psychology—cognitive-behavioral, Freudian-Jungian, and humanistic-transpersonal—are integrated within one Quantum Integral psychology and how this integral psychology gives us a much-improved vehicle for treating mental health.

  Book to read: Goswami, Shroff, and Onisor. *Quantum Integral Psychology and the Science of Mental Health*.

- Usually, until very recently, psychology's emphasis has been on abnormal psychology with the exception of transpersonal and depth psychologies developed in the nineteen sixties and seventies and positive psychology developed in the past decade. Alas! the new positive psychologies use different worldviews producing considerable consternation. You will learn an integrative approach to psychology using quantum science of consciousness and its experiences developed as a science of happiness. You can scale the ladder of happiness that quantum psychology predicts to increasing levels of happiness.

  Book to read: Goswami and Pattani. *Quantum Psychology and Science of happiness*.

- In the same vein, you will learn how to explore higher intelligence—emotional and supramental—to engage real life with all its infinite potentialities for which IQ intelligence is utterly inadequate. How to optimize your brain by integrating the brain and the body is part of this training as well.

  Book to read: Goswami and Onisor.
  The Awakening of Intelligence.

- You will explore the quintessential human quality of love between partners from the vantage point of a quantum activist. Using the theories of quantum physics and the discoveries in consciousness research, you will learn how to overcome your challenges in relationships for good. Unlike a quick fix superficial approach, you will delve deep and chart a path for yourself that can help you to eliminate obstacles like fear, manipulation and violence, leading you to true happiness in your relationships, turning the fictional "and-they-lived-happily-ever-after" into reality and unleash the power of love in your life.

  Book to read: Goswami and Onisor. *Love and the New Physics.*

- You will learn the quantum science of development from childhood to young adult, young adult through adulthood, and further development at and after the mid-life transition. This will help you to live to a ripe old age with quality of life intact. Having done so yourself, you will be able to guide people with *Quantum Life Coaching.*

- You will learn how to return the archetypes in human professions to not only help professionals to bring congruence in thinking, living, and livelihood but also enable the entire society to reenchant and transform to a quantum society.

  Book to read: Goswami and Onisor.
  The Return of the Archetypes.

- You will learn to explore quantum spirituality—spirituality in the service of the world—involving the embodiment of the archetype of wholeness ending in *quantum enlightenment.*

  Book to read. Goswami and Onisor.
  Quantum Spirituality.

- Ever since the discovery of the enchanted reality of ONENESS, explorers have wondered about *yoga*—how to integrate the oneness in our life. Great expositions have appeared on the subject: the *Bhagavad Gita, the yoga sutras of Patanjali,* Jesus' *Gospel according to Thomas,* Sri Aurobindo's *The Synthesis of Yoga,* but none has quite succeeded to take you the whole way. And now there is quantum yoga—the quantum path of practices that will take you to the promised land.

  Book to read: Goswami and Onisor.
  Quantum Yoga Manual.

- One of the great virtues of the quantumized higher education that we have developed for you is that it fully recognizes the importance of prosperity in your life in order for you to delve into the exploration of higher needs. To this end, we have developed the quantum extension of Adam Smith's concepts of capitalism. With this paradigm shift in economics, capitalism does not have to succumb to elitism, the economy can achieve full sustainability, and economics can serve both personal and societal interests because as we practice quantum style economics, our whole society becomes more and more enchanted reveling in our quantum connection.

  Book to read: Goswami. *Quantum Economics.*

- The implementation of quantum economics in our economic arena requires new business leaders and entrepreneurs. You will learn how to apply quantum principles of transformation for developing a new generation of quantum leaders in businesses.

> Book to read: Goswami, Sinha, and Goswami. *Quantum Leadership in Business.*

- We mentioned the idea of quantum higher education. Can these quantum principles be applied to lower education as well? The answer is affirmative, and you will explore the details of the new quantum paradigm of education.

> Book to read: Goswami and Onisor.
> *Quantum Education.*

- You will learn how to coach business leaders to quantimize their businesses, business entrepreneurs with their start-ups in subtle technologies, and business managers the science and art of emotional intelligence.

> Book to read: Goswami, Sinha, and Alvino. *Quantum economics and Business Coaching.*

If the program looks like a lot, don't be discouraged. The learning here is both conceptual and experiential. What you learn will be who you are and who you want to be. Learning is fundamentally different when it resonates with your being in this way. The complexity of the current higher education job training is that they try to fit you to a Procrustean bed—all size fit one. When you chart your own transformation, follow your own dharma, a learning life itself becomes enchanted.

INDEX

D

E

F

Lamarck 98, 99
left brain 136
liturgical 98

M

mantra 110, 149
material body 25, 100
material monism 16, 27, 32, 75, 91
maya 20, 23, 25
mental body 25
metaphysics 12, 88
midbrain 136
molecular biology 32, 33, 62, 67
monistic idealism 13, 17, 21, 22, 36, 38, 73, 80, 86, 88, 89, 134
monotheistic 22, 58, 59

N

Neel. Alexandra David 151
Neo-Darwinism 95, 96
Neuman, John von 18, 35
Nietzsche 6, 87
Nirvana 115
nonlocality 54, 55, 56, 61, 65, 66, 73, 77, 80, 82, 88, 89, 92, 97

O

oneness 13, 23, 26, 27, 36, 45, 47, 48, 50, 57, 58, 59, 60, 61, 62, 63, 66, 71, 73,
       79, 80, 81, 97, 114, 116, 176

P

pantheism 90
Physics of the Soul ii
polarization 10, 15, 27, 36, 56, 72, 75, 88, 113, 124, 131
prakriti 23
prana 24, 104, 108
promissory materialism 62
purusha 23

Q

qualia 50
qualified nondualism 21, 23
Quantum Activation ii
quantum activism 166, 171